Storehouses of Time

STOREHOUSES OF TIME

Historic Barns of the Northeast

Philip C. Ziegler

Down East Books • Camden, Maine

ISBN 0-89272-201-0
Library of Congress Catalog Card Number 84-51559

Design by Theodore R.F. Hall
Composition by Norma W. Young
Manufactured in the United States of America

5 4 3 2

Down East Books, Camden, Maine

To all of my children and theirs.
And a special dedication to Mr. Richard W. Babcock,
whose life is committed to the preservation of America's historic barns.

Contents

Phil
Ziegler
83

Preface

Soon after the idea for this study was conceived, I decided one day to draw my first barn for the book. Gathering up my sketching materials, I proceeded to a site where I knew a barn had stood for over 150 years. When I arrived, to my dismay I found nothing but the skeleton of my beautiful old barn standing amid a pile of rubble!

I sketched what was left of the barn anyway and then started to drive away. However, I found myself haunted by the memory of the old building and determined to join wholeheartedly in the battle to preserve and restore these storehouses of time.

To those of us whose lives are dedicated to the preservation of these venerable structures, barns are much, much more than just buildings to house tools, livestock, and feed. They are a very important part of our history. They are romance. They represent memories of the past. They are the guardians of our agricultural heritage.

The histories and dates of construction of the barns in this book are just as they were given to me, and although I have personally researched every possible detail, I make no claim to their absolute accuracy.

Many of the barns depicted stand today just as they were built on their original sites; others have been repaired, reconstructed, or moved to new locations. Many others are no longer standing but nevertheless deserve a place of honor in this book.

My illustrations are as accurate and true to detail as the Lord who first put a pencil in my hand has allowed me to draw them.

As I drew the illustrations for this book, my very soul seemed to become absorbed into the past. When I drew a foundation stone, my hands felt the grip of the chisel and the blow of the sledge. The chips flew as I constructed each hand-hewn beam, and as time passed, I grew to love these venerable structures and to respect their builders.

As each drawing came to life, I felt a sense of creation as I stood back, much as the original builders must have done, and with a sense of pride said, "It is finished, and may God preserve it for my children and theirs."

Acknowledgments

Writing this book has been a rewarding experience in many ways. The subject, historic barns, has been an elusive one, and without the help of literally dozens of dedicated people, the book could never have been produced.

I think back with joy to the many invitations I received from complete strangers to visit with them in their homes. I thank those who, although I was a complete stranger to them, trusted me with precious family photographs and documents because their dedication to the preservation of our agricultural heritage was greater than their fear that some of those records might be lost.

I am grateful to the directors of the various state historical commissions for giving me permission to use their archives to further the research for this book and to the many town clerks and city assessors who helped me to locate many barns.

A few people gave me more help than I ever expected, and I wish to give them a very special thanks. They are: Mr. Frank Beard, historian for the Maine Historic Preservation Commission; Mr. Walter Nebiker, Senior Historic Preservation Planner for the Rhode Island Historic Preservation Commission; Mr. James Parrish, Historic Preservation Planner for the Berkshire County (Massachusetts) Regional Planning Commission; and Charles Fisher, Scientist-Archeologist of the Peebles Island Division for Historic Preservation, State of New York.

Thanks are also due to Charles Klamkin, author of *Barns, Their History, Preservation, and Restoration,* and to Eric Arthur, author of *Barns,* for their permission to use photographs and other pertinent information from their books.

And I will ever be grateful to my mentor throughout the entire project, Richard Babcock, of Hancock, Massachusetts. Without his help I do not believe this book would ever have been written.

I also extend my thanks to John Markowitz, Mrs. Barbara Brainerd, Claude Cyr, Professor George Creeger at Wesleyan University, and the Webb family of Shelburne, Vermont.

Many other people were helpful in the writing of this book. They are too numerous to mention by name, but they have my sincere gratitude nonetheless.

1865

Introduction: Researching Barns

Often the first question that comes to mind as one explores the recesses (or even the ruins) of an old barn is, how old *is* it? That question naturally leads to others: Who built it? How was it constructed? How was it used, added to, and modified over the years?

About the only way to be absolutely sure just when a barn was built is to find a date stone (such as the 1865 date stone illustrated here), or a date carved somewhere in the timbers of the barn itself.

Without such information, a researcher will always have trouble pinning down the exact date any subject barn was built, but there are clues that any observer can train himself to notice. For example, were nails found on the site? Were they cut nails (sheared) or were they hand-wrought? Were treenails* used to hold the joints together? How were the beams joined? A barn with shouldered or splayed joints was constructed in the 1700s, while notched joints may indicate that the barn was built between 1800 and 1875.

Roof construction often offers valuable information. Are the eaves high or low? Are the rafters just tied or nailed together without the help of a ridge beam or board? Ridge boards and beams were brought into use about 1800. Prior to that time the rafters were simply fastened together at their apex.

Saw marks are other visible clues. If the marks are straight, it would suggest that the boards were cut in a saw pit with an up-and-down saw, indicating a construction date before 1800, while circular marks might date a barn as post-1800.

Determining the age of a barn and identifying how and for what purpose, it was built present a fascinating puzzle to those who like to study their history "in the field," so to speak. The illustrations and text that follow will, I hope, provide the solid foundation for developing an informed appreciation of a fascinating branch of our agricultural heritage. Readers who wish to learn even more about this classic architecture should consult the sources listed in the Bibliography section at the end of this book.

While exploring old barns, one cannot help but remark on their simple, functional beauty. It is hard to accept the fact that so many of them are on their way to total destruction. One day, a lovely barn is sitting there, and the next day there is a fast-food place or a condo or even a shopping mall. How sad! Developers should realize that, had they restored the old barn and built around it, the property would be far more attractive.

Why is so much attention diverted to the restoration of old homes and so little to agricultural structures? Much literature is printed extolling the virtues of old houses. Public and private money is freely spent on their restoration, while the historical value of a barn seems to be about that of a Christmas tree on December 26.

Perhaps this intense interest in historic houses is due to the fact that most of us are more knowledgeable about dwellings than about barns. The barns are just as valuable as sources of information about our past.

A special charm and mystery always surrounds these venerable structures. Every nook and cranny tells a story. The dark-stained wood in the corner indicates that a silo once stood there. A filled-in hole may show the presence of an inside well. Relics of the past abound if one looks for them. We must never let the day come when some future historian shows a picture of an old barn and says, "This *was* America."

*Treenails, pronounced (and sometimes even spelled) *trunnels,* were wooden pegs used in place of nails or spikes to fasten timbers together.

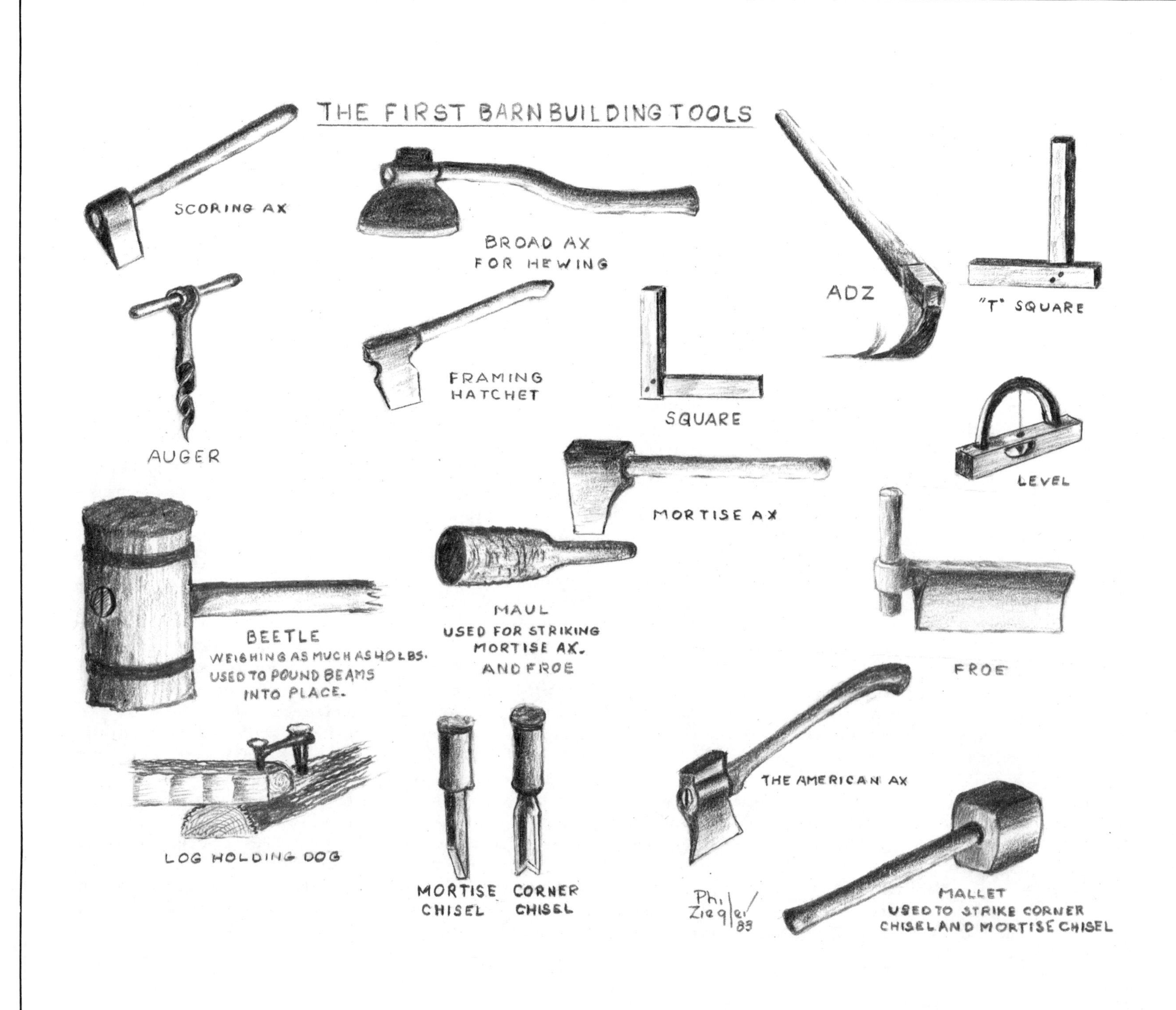
THE FIRST BARN BUILDING TOOLS
SCORING AX
BROAD AX
FOR HEWING
ADZ
"T" SQUARE
AUGER
FRAMING
HATCHET
SQUARE
LEVEL
MORTISE AX
BEETLE
WEIGHING AS MUCH AS 40 LBS.
USED TO POUND BEAMS
INTO PLACE.
MAUL
USED FOR STRIKING
MORTISE AX.
AND FROE
FROE
THE AMERICAN AX
LOG HOLDING DOG
MORTISE
CHISEL
CORNER
CHISEL
Phil
Ziegler
83
MALLET
USED TO STRIKE CORNER
CHISEL AND MORTISE CHISEL

The Builders and Their Tools

The earliest settlers in America found that foremost among their basic needs were shelters for themselves and for their animals, room for tools, and a place to store their food. Building materials were usually plentiful, but tools were scarce and of limited variety; therefore, these first structures were, of necessity, simple. The foundations were usually constructed of stone, the sides of rough-hewn logs, and the roofs of thatch or bark. The few pieces of manufactured furniture that the settlers had brought with them from Europe stood incongruously on hard-packed dirt floors, next to crude, handmade tables and benches hastily constructed in the New World.

These original structures usually served a dual purpose, with one part set aside as living quarters for the family and the rest for storage and livestock. The family would eventually construct and move into a separate cabin near the barn, but only those with ample means could afford such a convenience when they first started.

Larger homes and barns were built as their need became apparent. These barns were built to last and to meet the owner's needs; it is interesting that from Rhode Island to Maine, their basic structural parts were very much alike, even though architects were practically unheard of in connection with barns. There were no Greek columns or Gothic details in those early days.

However, no man could build a barn of any size by himself. He might cut the trees, hew the beams, and in general prepare the basic parts for assembly, but then his neighbors and possibly a master barn builder would come to his aid. After the ''bents'' were preassembled on flat ground, the foundation laid, and the sills properly placed, the ''raising'' began. No one was ever asked to do more than his share, but everyone — man, woman, and child — participated.

These people were not architects. They were not trained artisans. They were a determined group who were sure they would succeed because, deep in their hearts, many believed that God was there to help them every step of the way.

Very few trained craftsmen came with the first settlers from the Old World. Most artisans were protected by their individual guilds and respected for their talents, and therefore suffered much less than the average citizen from the religious persecution that caused many to emigrate.

Those who did make the journey often became leaders of the community. History shows that a blacksmith by the name of James Reed came to Jamestown with the first settlers in 1607, and it is reasonable to assume that his forge became a center for the bartering activity common at that time when hard cash was scarce.

This drawing is a composite that includes the basic equipment of many an early smithy.

A blacksmith's forge was his pride and joy. Stone more often than brick was used as the basic material from which a forge was built; this included the chimney and the hood above the white-hot charcoal. The base was usually a large, flat block of cut stone or masonry two or three feet thick. A hole covered by a square grate caught the ashes as they filtered down; later, these were removed through a side opening in the forge and often mixed with lye to make soap. Nothing was wasted.

Another hole entered from the back or side of the forge to accom-

modate a nipple that was cemented into the hole in such a way as to allow the outlet from the bellows to fit over it. This hole was placed just below the grate where the bed of coals rested, and the oxygen-rich air produced by the bellows heated the coals until they glowed white. It usually took an entire ox hide to make one set of bellows. This leather "lung" was attached to the leaves (wooden supports) of the bellows with large-headed nails made especially for that purpose. These leaves were about 1½ inches thick and from five to eight feet long, tapered at the forge end and terminating in a funnel.

As shown in the drawing, the top leaf of the bellows was anchored to a post by a strong board that passed through the post and was firmly attached to the top of the bellows. A flap built into the bottom board opened when pressure was released and closed when the bellows was activated. The shaft was fastened at its fulcrum point to a beam in the ceiling by a long steel rod. Another steel rod passed from the bottom board to one end of this shaft and a ring hung at the opposite end; the blacksmith pulled on this ring to compress the bellows.

The first priority for a blacksmith was the making of tools. A group of the most needed barn-building tools is illustrated. These were used mainly to form the beams necessary for most of the construction work.

Except for the American ax, which was an entirely American-developed tool, the rest of the tools illustrated are, more or less, self-explanatory. At first, this American ax, the most versatile tool made, had a long, narrow blade with a straight handle. Later, the curved handle evolved and it became even more popular, for this handle design kept the ax from slipping out of the user's hands. As steel was very scarce, the blade was, at first, made of plain iron. Even after steel became more available, it was still too expensive for the average user; as late as 1790, American axes were still made of plain iron.

When steel became more common — although still costly — a slit of pure steel was often inserted into the cutting portion of the ax by making the ax-head in two parts and then forging the cutting edge between the two halves. This edge, when sharpened, served almost as well as an ax made entirely of steel. Eventually the American ax was made of pure steel and became so popular that in the early 1700s English ironmongers often falsely labeled their axes "American" in order to sell them more readily.*

The prospective barn builder, after locating the proper trees, cut them to length with an ax. Then, using a compass (the kind you make circles with), he determined the size of the beam he wanted. This circle was then recorded somewhere in the barn and, when forming the rest of the beams, was used for reference. Measurements in inches and feet were rarely used when squaring off a beam except for the length needed.

*Edwin Tunis, *American Craftsmen,* page 20.

Next, the log was prepared for scoring by snapping a charcoal-powdered line at four places the length of the log. Then, with a scoring ax, it was cut crossways down to where the snapped lines formed a square. Anchoring the beam between other logs, the hewer would then stand atop it and, with his broad ax, chip off the excess wood down to the lines until a fairly flat side was formed. It was then turned so the hewer could stand on the flat surface and repeat the process until all four sides were hewn square. As illustrated, many broad axes had a handle bent outward in such a way that the hewer wouldn't cut his boot. Other hewers preferred just a plain straight handle. If a smoother beam was needed, another worker would use an adz to finish off the rough places left by the broad ax.

To make boards, a knot-free piece of pine was selected and, after squaring off the part to be used, the froe was placed at the top of the small, squared log and hit with a maul — thus splitting off a fairly nice, flat board.

Although the froe was useful for small jobs and short lengths of boards, the pit saw, or "up and down," was the mainstay of the board-producing industry for many decades. Although the pit saw was sometimes used to square up decorative beams, its primary function was the production of boards of varying lengths and widths. (The same saw was used for shaping roof timbers and studs.) Later, water-powered mills used the same up-and-down process until circular saws became prevalent after 1800.

Again, after the log had been squared a charcoal-dusted line was

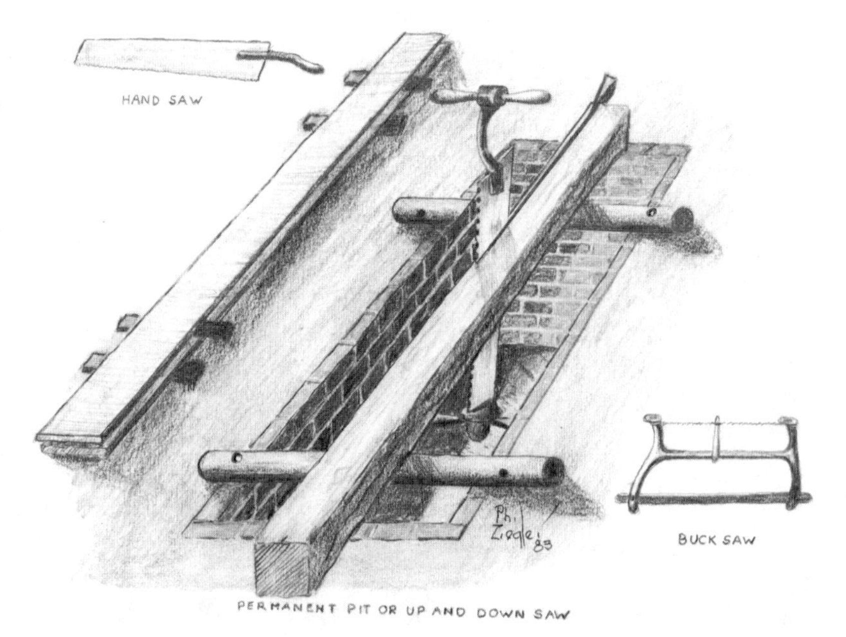

used to mark off the width of boards to be cut. The marked log was then placed on rollers above the pit. The top sawyer stood on it as he pulled the saw up; the man in the pit would pull it down, with the cut being made on the down pull.

Each cut was stopped about six inches from the end of the log and a new cut started because, should they have finished each cut, the top man would soon have been standing on air. After the cuts were made, the boards were cut to the desired length and stacked for drying, each layer separated from the next with scraps of wood.

Before the pit saw was perfected, in order to cut boards trestles were built above ground and the log was hoisted to the top of the trestle, thus necessitating lot of hard, back-breaking work. However, American ingenuity soon went to work and the idea of making a pit in the ground to take the place of the trestle was adopted. From just a hole in the ground, the pit evolved into quite an elaborate rig — brick lined, with brick sills laid level with the ground for the rollers. The job became even easier when one sawyer got the idea of swabbing linseed oil on the saw blades so the saw would cut much easier.

The next improvement was the "gang saw." This was a pit saw that had five or six saw blades rigged up in tandem, making it possible to cut several boards at one time.

The blacksmith's second priority after the making of tools was the production of nails. If he himself was too busy to make nails at any particular time, he relegated the job to his apprentice, as nail making was a continuing process.

Although nail-making machinery was available from Europe, it was so costly, and the nail-making process by hand so simple, that none of the machinery was imported to America for many decades.

The process of nail making began with appropriately-sized square rods that were procured from traveling ironmongers who, in turn, had purchased them from the iron-making mills that began dotting the countryside.* The ironmonger would travel from village to village selling many needed articles made of iron or steel; in short, he was a traveling hardware store.

The nail maker, whether a blacksmith or a farmer, would put the end of a rod in his forge, and after it became red hot, taper the end to a point. Each anvil had a "hardy," which was a hardened piece of steel about two to four inches wide, tapered to a sharp edge at the top (see illustration). This hardy could be easily removed when necessary. The craftsman would put a rod on the hardy at the desired length and notch it with his hammer, then put the tapered end of the rod in the appropriate hole in the anvil and break it off at the notch. The nail was

*In the late eighteenth century, as more and more iron-ore deposits were discovered and transportation became more available, dozens of iron mills were built around the Northeast.

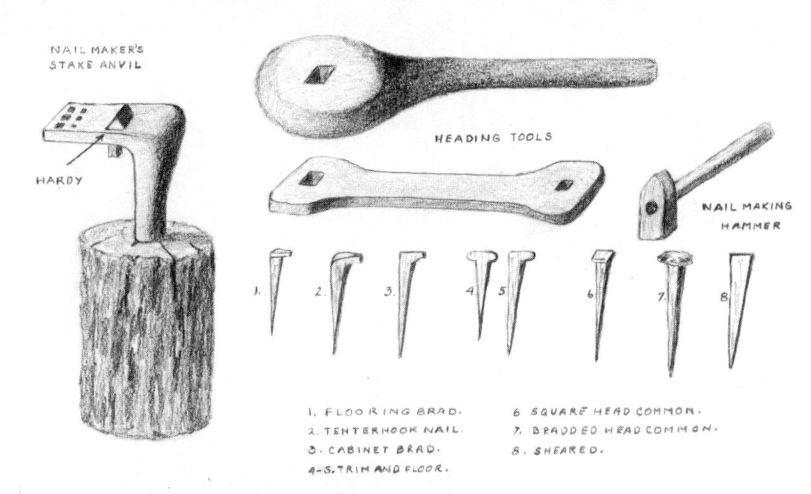

then headed with a few bradding taps of the hammer. A slight tap on the protruding end of the nail removed it from the anvil. Plain sheared nails were also available but these were of little use as they had no heads, and soon were abandoned.

Although the farmer with his limited equipment could make only the bradded-head common nails, the blacksmith was able to make the variety of nails illustrated here. Even so, almost every farmer's hearth served as a forge during the long winter months while the farmer made his supply of common nails. In spring he would take them into his village to barter for supplies.

Nails were priced according to size. Ten-penny nails were worth ten pennies per hundred, and so on. Today the designation refers only to the size of the nail, not the price.

Dutch Plantation Barn

Construction and Barn Raising

As the population grew, the production of food kept pace. The small, log-sided, thatch-roofed barn proved inadequate and a larger, more efficient structure began to take its place. These were still all-purpose barns, however, serving as both livestock shelters and food storage facilities; it wasn't until much later that the specialty barn came into being.

As each new barn was built, its style often betrayed its builder's nation of origin. If he was a Scot or Englishman by birth, his barns looked like those in his native Britain. If he was of Dutch or German descent, his barn took on an entirely different shape. These elevations show the fundamental differences in the frames of barns of varying ethnic constructions. Each builder brought his country's way of building barns with him, and yet, although no architects designed barns during this era, the basic principles of construction were the same from Rhode Island to Maine.

A good example of how the barns constructed by builders of different nationalities took on different shapes can be found in the large Dutch plantation barns, which were usually quite wide, with low side walls but high gable roofs.

Although there are many illustrations in this book of barns that look like they are ready to collapse, this appearance is often deceptive; these barns were built to last. Some of them may be restored to their original beauty where they stand, others moved to different locations, perhaps to become part of a museum or even as the framework for a house.

The German settlers were great barn builders. Their barns were generally much narrower and longer than the Dutch barns. Huge beams have been found in these barns as long as forty feet and measuring up to fourteen inches wide by twenty-four inches deep. Just imagine what a huge tree such a beam came from, and what a gigantic task it was to hew it into a beam of such proportions with the tools available at the time! (An excellent specimen of the German barn, located near Blenheim, New York, is described in the chapter on restoration.)

A very good example of a Dutch plantation barn can be found just south of Albany, New York. It has proved to be one of the largest Dutch plantation barns still standing. This barn, constructed *circa* 1800, boasts hand-hewn cross timbers two feet thick and over thirty feet long, and measures sixty feet wide by seventy feet long. At this time (1983), it is still standing, but it has degenerated too far for reconstruction to be feasible, and so is doomed to collapse.

This English barn, *circa* 1750, was found on its original site — the Madden Farm in Stephentown, New York. The barn was carefully

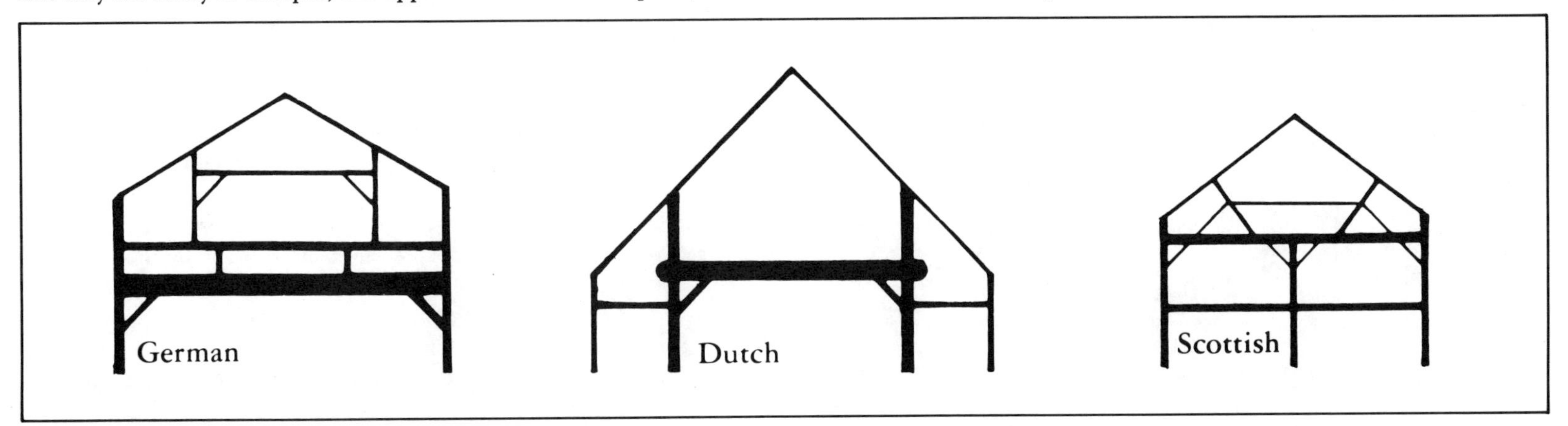

English Barn

dismantled and moved to Williamstown, Massachusetts, where it now stands as an integral part of a beautiful home.

When Richard Babcock was planning his barn museum at Hancock, Massachusetts, he wanted to have the typical Scots-Irish barn represented there. He found what he was looking for near Salem, New York. This thirty-two-foot by fifty-foot structure was built about 1750 by a true master craftsman from an early settlement. It is recorded that this barn was moved from its original site in the middle of Salem to the outskirts of the town because so much of the usable land was occupied by growing industries that there was no longer any use for this barn in the village. The illustration shows the barn as it looked before it was dismantled and transported to Hancock for the museum.

Scots-Irish Barn

The Babcock Barn Museum is now listed as a historical site on the National Register of Historic Places. Here, a haven has been established for many barns of varied ethnic origins; some are already reconstructed as they originally stood, examples of Richard Babcock's dedication to the preservation of our historic barns.

Trade guilds, or unions as they are called today, were formed in America from the time of first settlement up until the mid-nineteenth century, many of them in the building industry. Each builder left the mark of his guild in every barn he built — a circle, a triangle, or possibly a square. The Scots-Irish builders were prone to use secret joinery in their barns in the form of special locking dovetail joints or perhaps a distinctive corner brace. If one looks closely, signs of these guilds can be found in almost every barn. In one, for example, a cross timber and a wall plate are held in place with *six* treenails and *three* tenons and mortises.

In any type of construction, certain logical steps must be taken in their proper sequence, and any barn begins with the foundation. At first, these foundations were made of fieldstone, which was plentiful in most areas. As quarried stone became available, it too was used extensively, especially in the top layer of stone where a perfectly flat surface was desirable to support the sills evenly. These foundations were always tapered toward the top and were sometimes as wide as three feet at the base. The sills were then laid and the mortises carefully cut to receive the tenons on the bents. After prefabrication, the bents were raised, and the tenons fit into the designated mortises with uncanny accuracy.

A pamphlet about the Moon Barn at Williams College in Williamstown, Massachusetts, written by Richard Babcock, quotes one old-timer who remembered how the builders got the parts to fit so accurately: "That barn, the main part was put up with wooden pins. That's [the barn] been there a good many years, been put together with pins because they don't do that no more . . . haven't in a long time . . . . They would get together and get out those big logs and use a broad ax to square them. They would score the log and then they would come right back through and chop off the scored pieces and it'd be fairly square on that side. Then they'd turn it over and make chippings off the other three sides. And, after it was squared, they'd measure. Now, you'd be surprised at these old fellows that didn't have the education they should have, but they'd drill those holes when those beams were down on the ground and they'd get their pegs ready. Then they'd have a barn lifting, have the neighbors come in to help put these barns up, and I'll be damned if they didn't all come right to. They'd peg them right together."

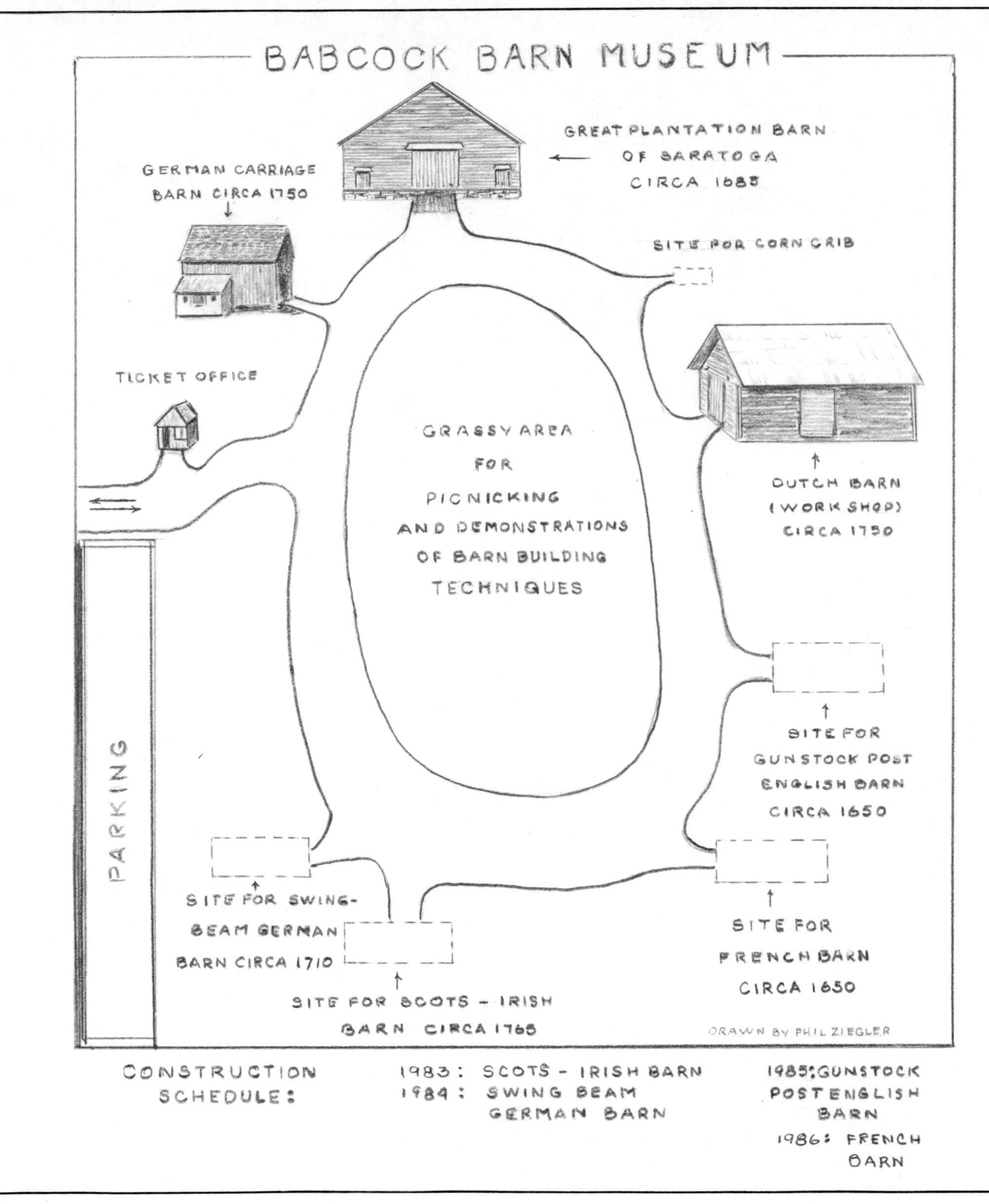
BABCOCK BARN MUSEUM
GREAT PLANTATION BARN OF SARATOGA CIRCA 1685
GERMAN CARRIAGE BARN CIRCA 1750
SITE FOR CORN CRIB
TICKET OFFICE
GRASSY AREA FOR PICNICKING AND DEMONSTRATIONS OF BARN BUILDING TECHNIQUES
DUTCH BARN (WORK SHOP) CIRCA 1750
SITE FOR GUNSTOCK POST ENGLISH BARN CIRCA 1650
PARKING
SITE FOR SWING-BEAM GERMAN BARN CIRCA 1710
SITE FOR FRENCH BARN CIRCA 1650
SITE FOR SCOTS - IRISH BARN CIRCA 1765
DRAWN BY PHIL ZIEGLER
CONSTRUCTION SCHEDULE:
1983: SCOTS - IRISH BARN
1984: SWING BEAM GERMAN BARN
1985: GUNSTOCK POST ENGLISH BARN
1986: FRENCH BARN

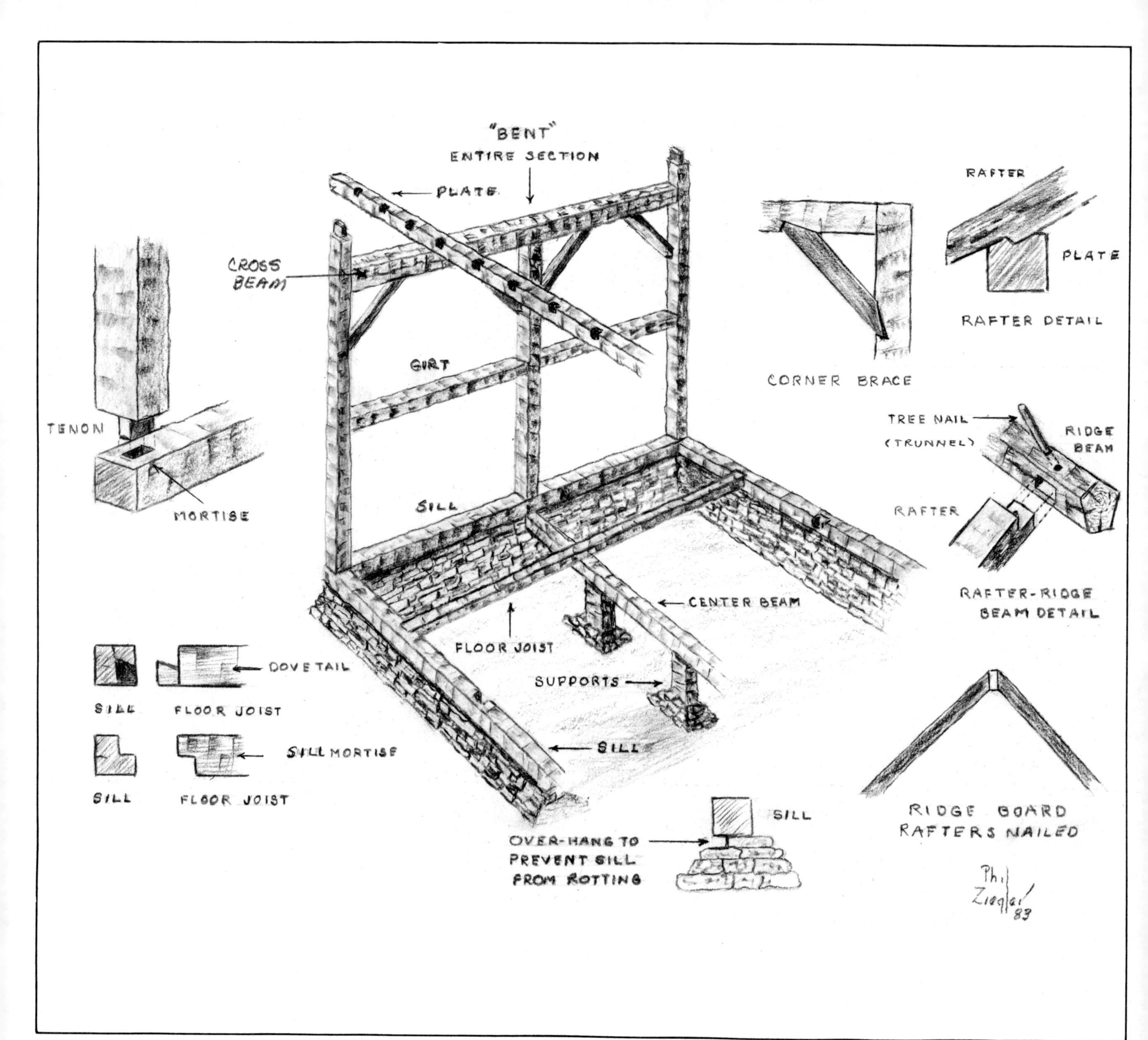
"BENT"
ENTIRE SECTION
PLATE
CROSS BEAM
GIRT
SILL
TENON
MORTISE
RAFTER
PLATE
RAFTER DETAIL
CORNER BRACE
TREE NAIL
(TRUNNEL)
RIDGE BEAM
RAFTER
RAFTER-RIDGE BEAM DETAIL
CENTER BEAM
FLOOR JOIST
DOVE TAIL
SILL
FLOOR JOIST
SUPPORTS
SILL MORTISE
SILL
FLOOR JOIST
SILL
SILL
OVER-HANG TO PREVENT SILL FROM ROTTING
RIDGE BOARD RAFTERS NAILED
Phil Ziegler 83

When prefabricating the bents on the ground, one workman used a beetle weighing as much as forty pounds to pound the mortises and tenons together while others secured them with treenails or wooden pegs. It must have been quite a sight to see these people all swarming around a lot of beams and then come up with a perfectly formed bent to put in its place when ready.

Before the bent was raised, the main girder down the center of the foundation had to be put into place and the cross-floor joists anchored. Enough floorboards were then put in place to give the barn raisers a place to stand when raising the bents.

If a barn was designed for dairy use, at this point the center sill was placed on top of the main center girder. Instead of using cross-floor joists, these sills were placed lengthways and joined to the end sills, usually with dovetail joints for extra stability, as were many cross-floor joists. A groove was routed out the full length of the sills to the depth of the floorboards that were to be used, so that the boards would fit flush with the top of the sills. All these boards were then put in permanently, except for the outside rows, the area where the cattle were kept in their stalls facing the center of the barn. (The old builders were smart. Livestock waste has a tendency to rot out the floorboards. When the boards needed replacing, all the farmer had to do was remove the rotted boards, which hadn't been secured, and replace them with new boards.)

Each master builder — English, Scots, Irish, Dutch, German, or whatever — used a distinctive type of joint in his barn construction. While English and Scots tenons were usually flush with the beam, the Dutch tenons protruded through the beams, taking various shapes — some rounded, some square, and others a combination of both. Many times, the shape of the protruding tenon depended on which guild the builder belonged to, and had nothing to do with the basic strength of the joined beams.

Sills supported the entire structure and kept it from getting out of plumb. Plate beams, the top beams running the length of the barn and tying the bents into one united framework, also helped to support the rafter beams. When an extra-wide barn was constructed, additional end sections were fabricated and tied into the crossbeams of the bents, thus creating more support for the larger, heavier roof. The lengthwise beams that tied these extra sections together were called purlins; they were notched to receive the rafters. They were installed about halfway between the plate and the ridge board or ridge beam.

In the next drawing, the loft is drawn with sawn boards forming the floor, but only the farmers with ample means could afford such a floor; the average barn loft was floored by simply laying poles together. However, perhaps a pole floor was more efficient and safe than boards, as they allowed better ventilation of the hay piles stored in the loft.

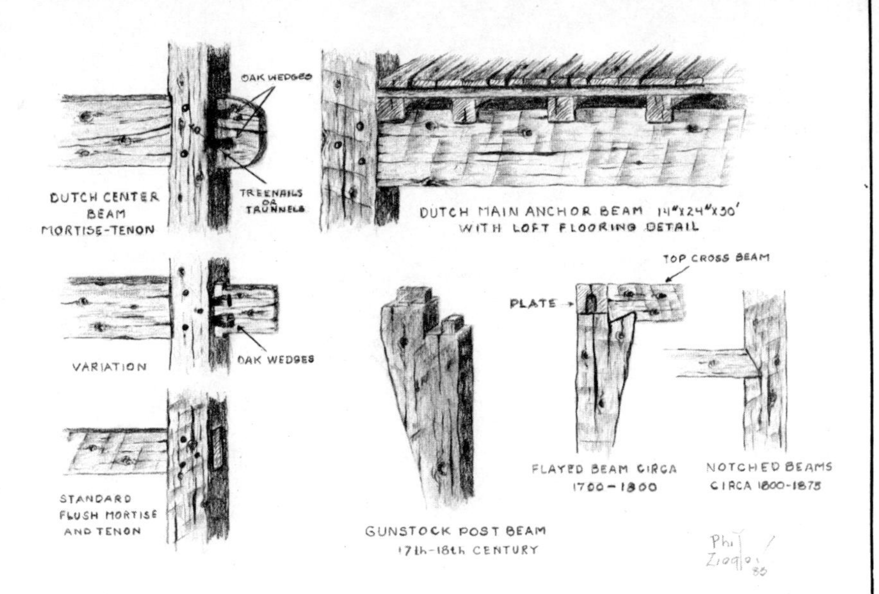

As distinct from carpentry, in which, after 1860, the builders used sawn boards and joists, joinery was a trade by itself. Pride in his craft caused the master builder to scorn such uncouth methods and materials; sawn joists seem to have been forbidden, and tenons and mortises were made by hand to strict specifications. They were so accurate that in many cases, where a series of barns was built by one master builder, parts of one barn were interchangeable with like parts in another. Mr. Babcock tells of a case where one of the beams in a

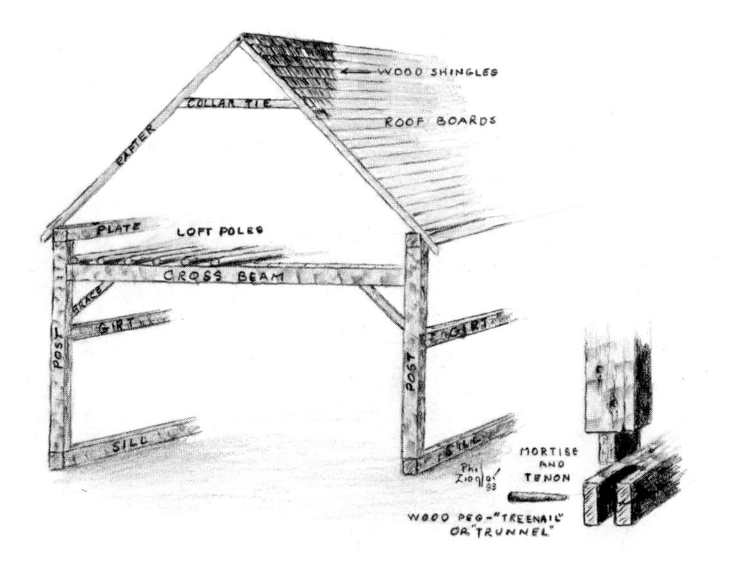

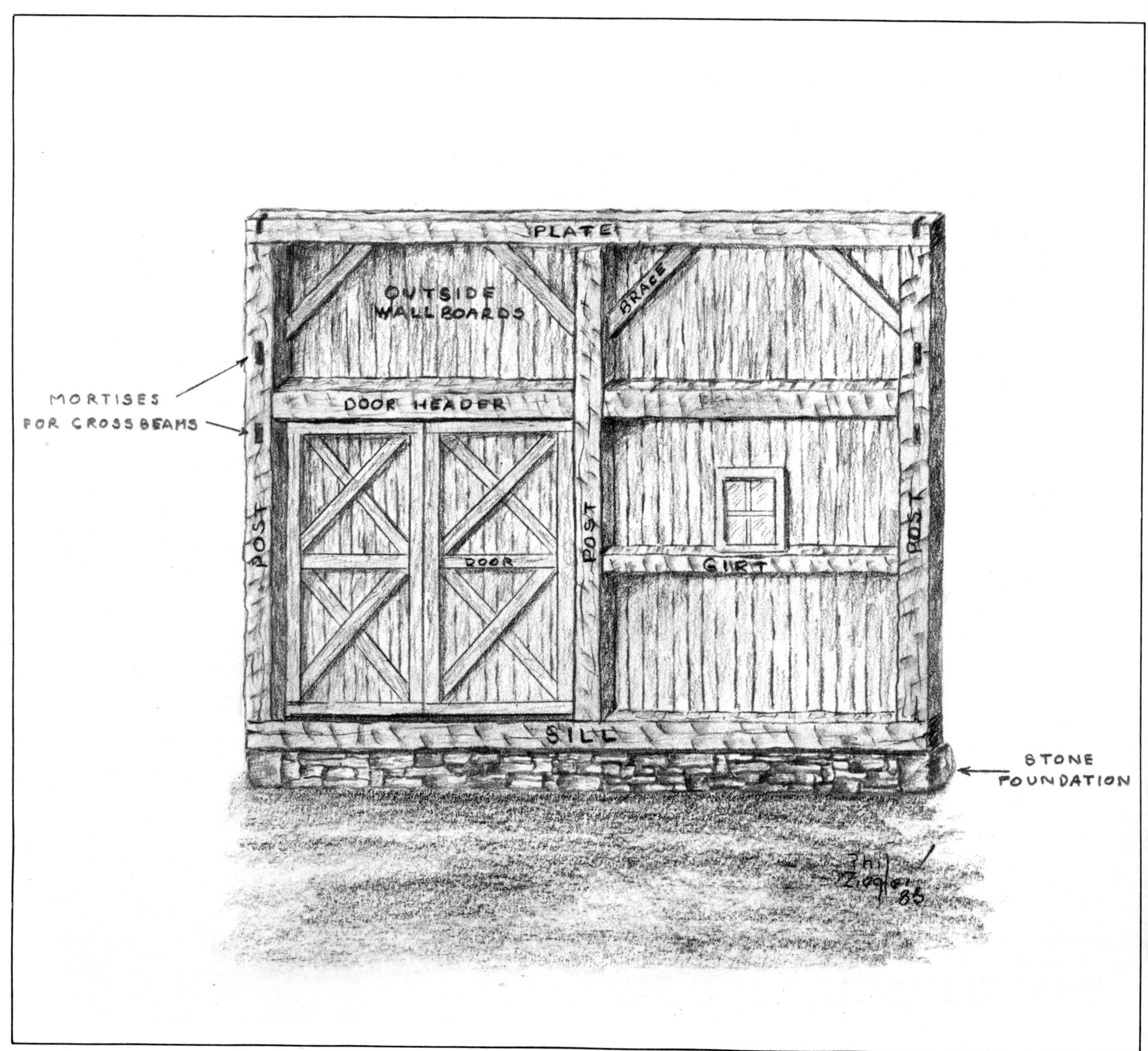
PLATE
OUTSIDE WALLBOARDS
BRACE
MORTISES FOR CROSSBEAMS
DOOR HEADER
POST
DOOR
POST
GIRT
POST
SILL
STONE FOUNDATION

barn he was readying for reconstruction was so rotted that it was useless. Knowing that a barn he had recently located was constructed by the same builder, he went to it, looked it over, and, sure enough, spotted a beam that appeared similar to the rotted one. He had it transported to the reconstructed barn, where it fit perfectly. That is precision indeed, especially considering that no machinery was involved and that only a compass circle in one barn was the basis for all of the builder's mathematical calculations. This proves that prefabrication is not such a modern innovation as one might assume.

A barn raising was to a barn builder what a quilting bee is to a quilt maker: an opportunity to combine work and pleasure. After the foundation stones, beams, and sills were all in place, the bents preassembled, and a prayer spoken, everyone — some fortified with ale — joined in the fun.

How did a group of people, regardless of how strong they were, move partitions (bents) weighing many tons into place? And how did they handle some crossbeams weighing as much as ten thousand pounds? Like master builders the world over, they called on the physical principles described by the Greek mathematician Archimedes, who made many discoveries concerning leverage and specific gravity and who designed such lifting and pulling devices as the capstan (or as the barn builders called it, the bull wheel) and a type of derrick known as the gin pole.

Actually, there were three ways a barn could be raised. The first was the hand-raising method. This consisted of everybody pushing and swearing together (swearing was, no doubt, permitted during barn raisings), lifting the bent as far as they could by hand. Then men with pikes — poles about sixteen feet long with a zig-zagged sharp point — took over. These pike men jabbed their hooks into the bent and began to raise it after the hand raisers had lifted as far as they could. With anchor men in position to guide the tenons into their proper mortises, using a stout, heavy pole, the bent was finally raised. At this point, it was secured and aligned by nailing a large board to both the bent and the sill, forming a secure brace. The treenails were pounded in, and the crew was now ready for the next bent; the whole process was repeated until all the cross sections were in place.

This drawing is of an actual hand raising done at Framingham, Massachusetts, in 1982.

A second method of raising a barn was hand raising using a gin pole. Deferring again to Archimedes, farmers used the gin pole for much larger barns. With larger and heavier bents and beams being used, hand raising by itself was no longer feasible.

The gin pole was actually a strong young tree of proper size and length that, after trimming, was then secured to a cross log at the bottom with two heavy braces. This formed a T at the bottom, and the gin

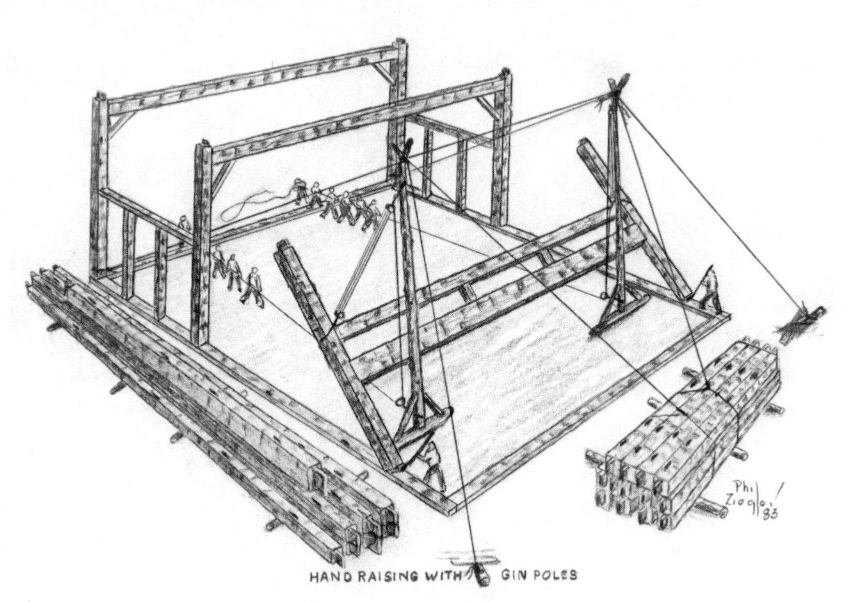

HAND RAISING WITH GIN POLES

pole was ready to use. As the drawing shows, the gin pole was left with a V or two short, trimmed branches at the top to which all ropes and pulleys were secured; the V prevented the ropes from slipping down the pole. Guide ropes, usually three, were then tied to the top of the gin pole and anchored to a tree, a stake driven into the ground, or even just a pile of beams. A rope was then strung through a pulley tied to the top of the gin pole, run through another pulley at the bottom, and out to

BULL WHEEL

HAND RAISING AT FRAMINGHAM, MASS.

the men doing the pulling. Then the tug-of-war began. As the bent was raised, the anchor men again did their work, as in a hand raising. When the bents were all in place and side-braced, a single long beam called a plate was then joined to the top of the upright end-beams on the bents and fitted to the precut tenons and mortises. This beam was notched at intervals to receive the rafters. The skeleton was then ready to have its "skin" put on and a roof to top it all off. This illustration shows two gin poles being used because of the size of the bents.

The third and most efficient method of raising a barn was the gin pole and bull wheel combination. The bull wheel is similar to the capstan used to raise the anchor on a ship. This combination proved efficient enough to raise even the heaviest bents and beams. It took only three men to operate a bull wheel: one on each end of the pole that went through the post, and a third handling the "out" rope. The next

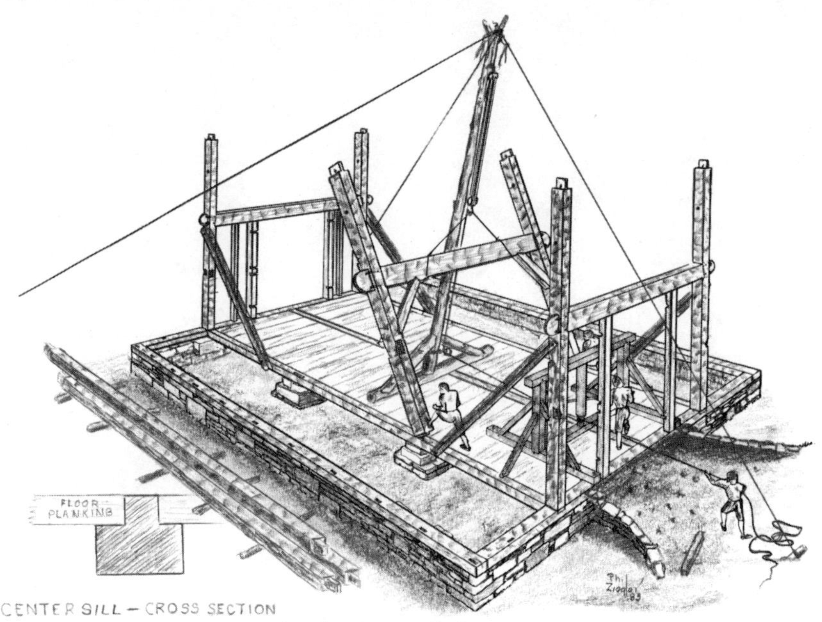

CENTER SILL — CROSS SECTION

GIN-POLE - BULL-WHEEL RAISING

illustration is a drawing from life of Mr. Babcock handling the out rope on a recent reconstruction, the Phillipsburg raising at Tarrytown, New York. This is also referred to as the "Sleepy Hollow" restoration.

The barn in this illustration, showing the whole gin pole/bull wheel process, is destined to be a dairy barn, indicated by the center sill running the entire length of the barn, as opposed to an all-purpose barn, where the floor beams run from side to side. The cross-section shows the routed-out sill, described earlier in the chapter.

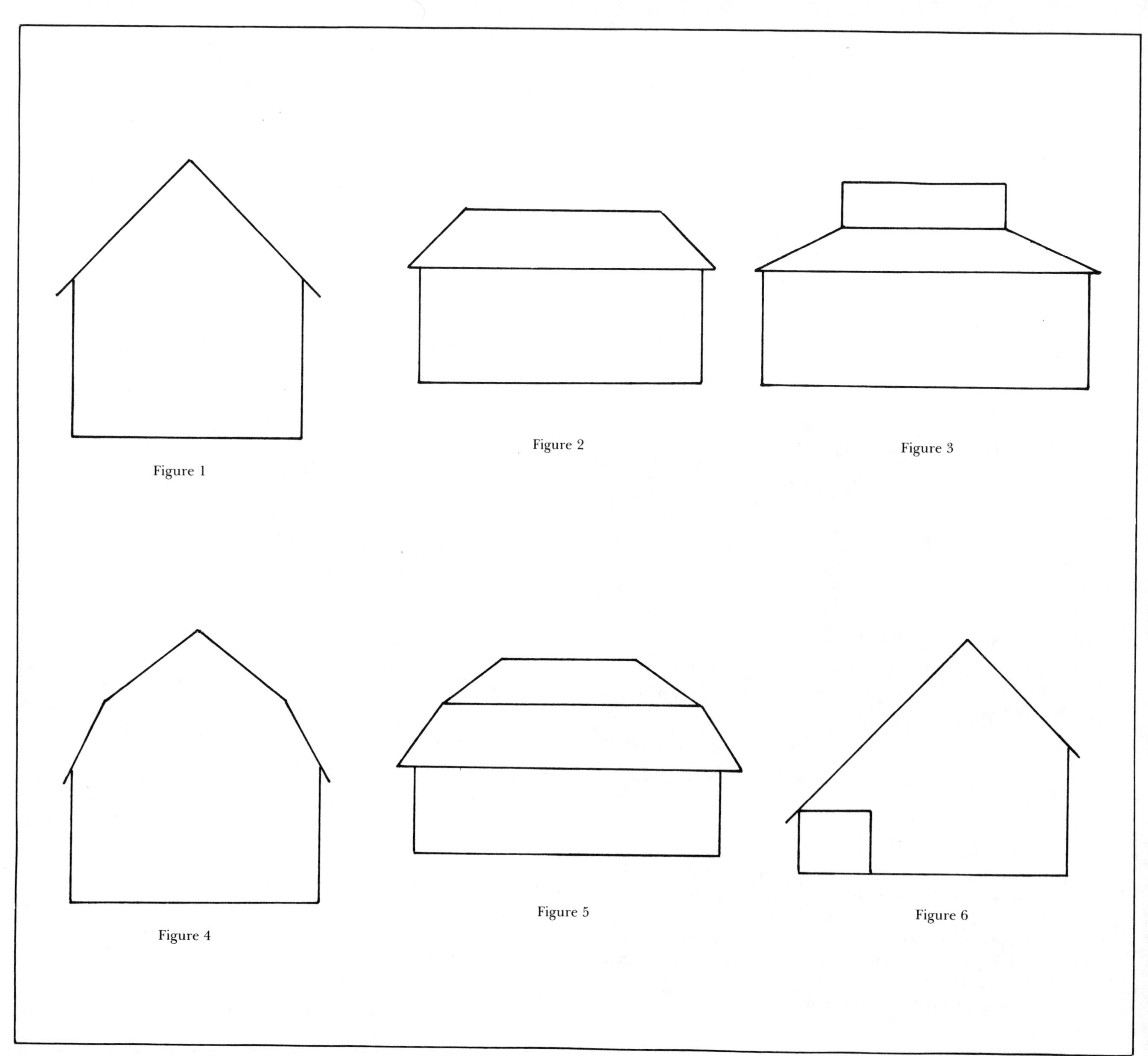

Figure 1
Figure 2
Figure 3
Figure 4
Figure 5
Figure 6

Roofs, Silos, and Cupolas

ROOFS

A critical part of barn building, second only to the framework, is the construction of the roof. Through the years, many different designs have appeared, from the simplest, the gable roof, to the most sophisticated, the mansard.

Roofing materials also changed as the decades passed, from the earliest thatch, to the most modern, the "tin" roof, to the most expensive, the copper-sheathed roof. (Although tile has been commonly used in roofing since Roman times, it is not found on barns in the Northeast.)

The gable roof, Figure 1, occurs when the triangular end-walls of a barn terminate in the roof. A perfect example of this type of roof can be seen in the James Smith Homestead, located in Kennebunk, Maine. Built in 1753, it is a well-preserved rural complex in the Georgian style. Southeast of the house is this large, gable-roofed, two-story barn, built about 1880, which has shingled sides and narrow clapboards on the ends. The end facing the house contains a large sliding door with a noteworthy ornate pattern. A wing with a shed roof is attached to the northwest side of the barn, and a secondary doorway faces east.

This Georgian complex dates from a period of marked prosperity that occurred in Maine between 1750 and 1760. Most of the villages in the area had, like Kennebunk, been depopulated in the Indian uprisings of 1690 and after. Resettlement of the area began about 1715, but for a number of years growth was very slow. However, good farming land was especially sought after by daring, energetic men such as James Smith of York, Maine, who established this fifty-acre homestead in the Kennebunk area.

Figure 2 shows the hip roof, *circa* 1760. This is a variation of the gable roof, formed by slanting the end section in from the eave at much the same pitch as the long section of the roof. It is also sometimes referred to as a cottage roof.

The hipped gable roof, Figure 3, is another variation of the gable roof, and Figure 4 shows the gambrel roof, which, like the gable roof, is supported on the ends by the vertical wall. However, instead of being totally on the same plane, the bottom section angles steeply from the eave to about halfway up. From that point, the rest of the roof angles to the peak. The bottom section was either curved or straight; it all depended on the whim of the builder.

This roof design was developed around 1750 to increase the storage capacity of a barn under the eave section, and because of this advantage, it soon became quite common.

A perfect example of this type of roof is found in this drawing of Norlands, located in Livermore, Maine. Norlands, the Washburn family estate, is a remarkable survival from the nineteenth century. A school, church, house with attached barn, and library stand amidst two hundred acres of beautiful surroundings.

In 1843, Israel Washburn, Sr., tore down his first house, built about 1805, and replaced it with one somewhat in the Greek revival style. He decided in 1865 to remodel both house and barn and in June 1867, the gambrel roof on the barn was completed. Later the same month, unfortunately, the buildings were destroyed by fire. The Washburns immediately created another home; while the ashes were still cooling, a new complex was being designed by George W. Thompson. Even as he worked, his men were in the nearby woods, already cutting wood for the project. The basic work on the new home was completed before the year was out, and in 1868, the house and barn were both ready for occupation. In 1915, extensive repairs were done to the 1867 barn; five years later the barn was replaced with a massive gambrel-roofed one. The Washburns boasted of having the finest barn in the area, claiming that one hundred tons of hay could be stored in the lofts.

In the decades since 1920 no other important alterations have been made and the illustration shows the barn as it stands today.

Francois Mansard, the French architect, is credited with the design of the mansard roof, Figure 5. It is really a combination of the hip gable roof and the gambrel roof. Although it was a delight to look at, it was a builder's nightmare, and only the richest farmer could indulge his fantasies by building such a barn.

Only one mansard-roofed barn, the George Wise Barn in Kennebunk, Maine, could be found to represent the style in this book. This barn, built about 1868, is on the site of an earlier one built or bought by Daniel Wise in 1775. The main house was moved in 1868. The present house and barn were built by George Wise, son of Daniel, and there George died in 1892 at the age of eighty-two. This handsome dwelling and barn are the finest in the district, and the mansard-roofed barn is crowned with a beautiful cupola, unusual in that it boasts a wire-fenced perimeter.

Figure 6, the saltbox roof, is the only one of original American design. It appeared on the agricultural scene about 1850, the same time as the gambrel roof, evolving when farmers began to need extra space for their fast-growing inventory of machinery. By extending one side of the roof over a shed built up to the barn, extra room was created. The general assumption is that all of these roofs on the low side faced the direction of the prevailing winds, but in most cases, the low side faced directly north.

In the earliest years, some roofs were thatched, others sodded. The sodded roofs were usually the work of Irish settlers. Thatch or sod roofs were a necessity at first because rafters were usually just poles fastened together at their apex with other poles running crosswise to form a base for the roofing material. Later, rough boards replaced the horizontal poles, forming a base for solid roofs covered with tree bark. The bark was laid in rows, one with the rough bark side up and the next with the smooth inside uppermost, making an interesting, alternative pattern. As more and better boards became available and center-beam construction was adopted, shingles took the place of other materials. As the decades slipped by, more alternate materials were used, such as asphalt shingles, slates, and aluminum or steel.

The center beam eventually was superseded by a center board at the roof apex. Rafters were attached to the center board with long nails, thus eliminating the need for mortise-and-tenon and treenail joints.

The George Wise Barn

SILOS

The first upright silo was built in McHenry County, Illinois, in 1873. Travelers returning from that area brought back news of this marvel, but it was not until twenty years later that these practical structures were introduced into the Northeast.

In the beginning, the silo was a square wooden structure built inside the barn in one of the corners. One can tell by the blackened stains in the wooden interior just where such a silo once was located because the damp silage always stained the wood it lay against.

Next, silos appeared in a circular form but still inside the barn, dead center. Two such examples stand out: the great stone Shaker round barn in Hancock, Massachusetts, and the round barn in Shrewsbury, Massachusetts. Illustrations of both of these barns can be found in the chapter on circular barns.

By the mid-eighteenth century, the charcoal industry was beginning to use up much of the available wood to fill the maws of the new, fast-growing iron foundries. As a result, other building materials besides wood were being investigated. These included stone, brick, and even, at times, poured concrete. Silos constructed of many of these sturdy materials remain standing today. A beautiful, tall stone silo nestled against a well-constructed barn is indeed a picturesque sight.

The silo was eventually moved to the outside of the barn, and the reason lies in why they were built in the first place: as a storage place for silage, which is green fodder. Although silage should be stored in an airtight place, many farmers who could not afford to build such a silo made theirs of wooden slats bound together, row after row, by a steel band or heavy wire — much as a barrel is made. Also, silos were very hard to clean inside and were a source of some serious fires, a subject that will be touched on again later in this book. And finally, because these wooden silos were not airtight, many a crop of fodder was lost through spoilage.

In the airtight group of silos, stone was the predominant building material and, at first, the silo was as deep in the ground as it was above it. Then farmers began to realize that a stone silo with three-foot-thick walls had very little chance of ever falling over, so they began to build them entirely above ground.

As longer wooden boards were turned out by the lumber mills, silos became less expensive and easier to build. Although such silos were not as airtight as their stone and brick counterparts, they turned out to be fairly airtight if carefully constructed.

When outside silos first began to appear they were often made of very short pieces of board, bound together as tightly as the builder could make them; those silos, however, became the victims of every prevailing wind. It wasn't an unusual sight to see several of them on

the horizon all leaning in the same direction, pushed there by the dominant gale. The next illustration depicts one such silo as it leaned like the Tower of Pisa beside its barn, which is still in excellent condition.

Silos made of brick and stone came in all shapes and sizes, while those made of poured concrete were usually round. From left to right, the next illustration shows a wooden round silo at the left. In the center is a round one with support ribs from top to bottom made entirely of

ROUND WOOD SILO — ROUND RED BRICK SILO — OCTAGON STONE SILO

red brick. The last is made of flat, fitted fieldstone. The tops are missing on the last two but the first one shows that even on silos, the farmers ''fancied up'' their farms by making the roofs into cupolas. (An interesting side note: One night, while traveling through northern New York around Christmas time, I saw silos on several farms, each with a beautifully decorated Christmas tree perched atop the roof and a myriad of colored lights twinkling merrily away.)

CUPOLAS

Very little is known about the origination of the cupola, but variations of it are found on barns throughout the civilized world. Originally, the cupola was built as a decoration on a barn or house and was constructed directly upon the roof. It gave the building a unique feature; one could say that a cupola was the product of the builder's vanity.

These cupolas were to be found on many barns, and not until fires caused by the internal combustion of the hay stored in them began to take a fearful toll did the idea of ventilation enter the farmers' heads. Barns were consumed at an alarming rate, and the owners began to seek ways to eliminate this hazard. Good ventilation proved to be part of the answer, and soon just about every kind of opening imaginable was put in the barns. In stone barns, openings were pounded out of the sides to form vertical or horizontal slits; eventually these were built into the barn as it was constructed. Sections of walls in wooden barns were made to open and shut with the aid of ropes and pulleys. Windows were put in and openings were cut under existing cupolas. When new barns were built, a hole was designed in the roof to accommodate the cupola, which had windows, louvers, or other openings.

The early cupolas of the ventilating type were of plain louvered-window construction, such as the one shown at the top left of the accompanying illustration. As time passed, they became more decorative and their variety was limited only by the imagination of the builder. Various types of cupolas are shown throughout this book.

Barns of Rhode Island

The Gardiner-Arnold Barn, built about 1870, and located on Scrabbletown Road in Kingston, was the home of J.S. Arnold until the late nineteenth century. The South County Museum occupied the building from the late 1930s to 1977.

This fine, gambrel-roofed structure still stands across the road from the old homestead. In addition to the barn, the group includes two large utility sheds and a milk house. Two family cemeteries remain on the property, with some gravestones dating back to the eighteenth century. It is unclear whether Joseph S. Arnold, John Arnold, or Sylvester Gardiner actually built the barn.

A farmhouse and a very interesting group of outbuildings on a working farm in the northeast corner of Burrillville comprise the Mowry-Wright farm. The large farmhouse and barn were probably built in the eighteenth century. In the early twentieth century the property was known as the Ironstone Country Club.

Around the house are some high stone walls, and nearby is the wood-shingled farm complex. It is sited on a hill and dominated by the large barn, which boasts a pair of cupolas.

The gable roof is covered by hand-split shingles (thought to be the original roofing). It is of post-and-beam construction, with mortise-and-tenon joints fastened by treenails. The entire framing is in excellent condition though the exterior needs much work. No renovation work is planned for the near future, however.

The Henry S. Nichols Farm is on Tarkiln Road near Burrillville. The barn is a large, 1½-story building with narrow clapboard siding, a wood-shingled gable roof, bracketed cornices, and two large barn-door openings. The cornices on the barn complement the cornices on the house. This *circa* 1860 barn is no longer being maintained, so it may soon become just another pile of barn rubble.

Phil Ziegler 83

This lovely old barn and attached wagon shed, built about 1810, is located in Exeter, on the Simon Lillibridge farm. It stands about 150 feet from the farm house between two vegetable and flower gardens set within a stone-walled enclosure.

This barn is the architectural focal point of the Lillibridge farm complex, now owned by Peter and Margaret Lacouture. It is a gable-roofed, two-story, frame and granite structure set flank to the road. A full-depth, one-story open shed, closed in at the southeast corner, is on the south side, and a pent-roofed bunk room is on the northeast. These additions vary the rectangular shape of the main barn, which was originally of quite traditional construction and layout.

About a fifth of the ridge pole is missing and extra uprights have been added above the main level running to the roof peak. Also, some diagonal corner braces have been replaced, but by and large, the original hand-hewn framing remains intact.

At a right angle to the barn is a separate three-bay wagon shed set on a three-foot granite foundation; this appears to have been built in the early nineteenth century about the same time as the barn, since the hand-hewn framing and granite work closely resemble those of the main structure. The whole complex is sided with exposed vertical-plank sheathing, except for the gable end of the section over the shed, which is sheathed with hand-split shingles, as is the entire roof, including the shed.

''The Stone Barn'' is not a very romantic name, but that is what this barn is called. It is located on the Stone Barn Road in Burrillville. Built about 1855, it is a rubble-work building with a large door opening. Several smaller doors and a row of little rectangular openings across the top of the large doors make it a truly well-ventilated barn.

Stone barns are very rare in Rhode Island; this particular one was built by an Englishman and carries a bit of the flavor of merry old England in its lines, reminding viewers of an English country cottage.

The former William Steere farm in Gloucester is a fine example of an agricultural complex including a farmhouse and barns, an orchard, and some beautiful farmland, set off by a picturesque stone wall.

Built about 1780, the house is a 2½-story, end-gable structure with a large, brick central chimney. The entry is near the center of an asymmetrical, four-bay, south-facing facade.

This drawing shows the larger of two vertical-board-sided barns across the road from the house. The original builder spent a lot of time on details in framing this barn, evident in its excellent condition inside and in the striking hand-hewn beams. The pit-saw grooves are still visible throughout the barn on the rough-cut boards.

Although not in the strictest sense a barn, the Daniel Cooper Carriage Shed (*circa* 1820) is included in this book because it represents a perfect example of this type of building. Its internal structure of hewn beams, treenails, and pit-sawn boards is the same as any barn, only smaller. Named after its builder, it is located near Burrillville.

No doubt this carriage shed has an interesting history, but no further details about its use or former owners are available.

Located on Lewiston Avenue in Richmond, this *circa* 1680 barn stands near the 1½-story, gambrel-roofed Samuel Clark house. The farm includes a one-room schoolhouse, a machine shop, and this late-Victorian, shingled barn with its lovely cupola. A well-maintained family cemetery is located nearby.

The Steven C. Brown Barn, located in Foster, was built about 1850 (no exact date can be determined). It appears to have been built in two stages, and an unusual feature graces the post-and-beam interior: the inside wall of about half the barn is covered with wooden shingles. A crawlspace underneath the barn slants from about one foot to four feet in depth. When the vertical boards were put on the end of the barn, small spaces were deliberately left between them in order to create more ventilation for the haymow.

The barn was primarily used for raising chickens. It is now owned by Steven C. Brown, who purchased the property in 1963. The house was built in 1908 and acquired an addition about fifty years later.

The Randall farm, in Foster, was originally settled in 1780 by William Randall; his sons Zephaniah and Eddy divided it in 1819. The early nineteenth-century barn is in excellent condition. There has been only one major improvement made: the recent addition of asphalt shingles on the roof. Vertical barn boards face the sides, while shingles cover the ends. A stone-lined entrance dominates one end and leads to the area under the first floor.

This beautifully preserved barn, *circa* 1785, stands on the Paine-Ross farm on Paine Road in Foster. About the only concessions toward renovations are a modern shingle roof added in 1950, and periodic coats of yellow paint.

Structural clues indicate that the right-hand half of the barn, where the sliding doors are located, was built some years before the rest. The framing in the older section consists entirely of hand-hewn beams pegged with treenails and auxiliary boards cut with the up-and-down saw, whereas some circular-saw evidence can be found in the newer portion. The foundation is built entirely of stone. This new part was added in 1835; it also has hand-hewn beams and traditional oak-treenail pegging. Some structural repairs were made after the 1935 hurricane but nothing new was added. Some sill work now needs to be done as the barn appears to be settling in certain sections.

Phil
Ziegler
'83

Barns of Massachusetts

The Henry J. Harlow farm was originally called Echo Farm, but that name has not been used for many years. It is owned by Henry J. Harlow of Shrewsbury, whose great-great grandfather purchased the original property in 1796. Mr. Harlow's niece, Mrs. Barbara Brainerd, was very helpful in locating this barn.

Family lore has it that Thomas Harlow, an apprentice tanner, and his father rode to Shrewsbury from Duxbury one fine day with one thousand Spanish silver dollars in their saddlebags. They purchased the property upon which this barn was eventually built, and Thomas set himself up in the tannery business. He was very successful until he discontinued the business about 1845.

The barn shown here was built in 1859 from parts of another old barn on the property, leaving about one third of that original barn still standing. The 1859 barn is still in excellent condition, much the same as when it was originally constructed. A large, square wooden silo once stood to the left of the barn but it has been gone a long time. A milk shed now occupies the spot.

On one of the top braces of the Oxbridge English Barn is carved the date it was built: 1715. It is located in Oxbridge, and is undoubtedly one of the first barns to appear in the area of the Blackstone River at Oxbridge. According to one of the village historians, the house was built in 1720. This indicates that the owners probably lived in the barn until they could manage to construct a separate dwelling place for themselves. Another source interviewed insisted that he was absolutely sure that the family lived in a small log cabin adjacent to the barn when the barn was built.

This barn is shown in the very early stages of its reconstruction by Richard Babcock, of Hancock, Massachusetts. The post-and-beam framing is still in excellent condition, but the original up-and-down-sawn siding will be replaced during renovation.

The Francis Clark Barn, *circa* 1773, was built by a settler who came to Tyringham with his wife Mary (Johnson) from Middleton, Connecticut. Clark dealt in dairy cattle, horses, and livestock, and was a noted trader in the area. He died in 1813, and the farm passed down through several generations of the Johnson family until it was purchased in 1946 from a descendant, Annette Hilliary, by Bill and Marion Gelschleichter.

The round window in the gable was reconstructed at that time from parts of the original window. The vertical-board ell was built by Stephen Johnson in 1872 as a tobacco shed. (Tyringham is one of the few areas where tobacco was ever grown in the Berkshires.) One of the addition's support posts rests on a stone inscribed "S.C. Johnson," and under the name is a tobacco plant carved in bas-relief.

The darker square area shown on the broad side of the barn is where a window or hay-loading door used to be.

Elisha Cook and his son built this barn in about 1840, in Orleans. The property remained in the Cook family until 1921.

Most of the records on the Barley Neck Farm were destroyed in a fire that burned down the Orleans Town Hall in 1862, but the information still available indicates that it was a very busy farm. Originally used strictly as a dairy barn, with hay in the loft and cows installed on the threshing floor, the building was later (up to 1950) used mainly for raising cows and hogs. The present owners, Raymond and Barbara Tannuzzo, are in the process of renovating the original house and barn; the remainder of the farm was subdivided into building lots in the 1970s.

The Doane family built the Mayo Barn in 1820 and sold it to Benjamin Mayo in 1868. It is located in East Orleans. This barn was used continually until 1960 to house livestock and to store grain and hay. The principal crop was poultry — ducks and chickens. In 1910, additions were built on. The original barn is that portion under the pitched roof. The entire group of buildings has shingle siding.

Although the Stephen Phillips Barn doesn't appear to be especially large in this drawing, it is actually one hundred feet long, forty feet wide, and sixty feet high at the ridgepole. It was built by Stephen Phillips in 1827, sold in 1838 to a Charles Lawrence, and is presently owned by the St. John's Preparatory School in Danvers. It is a very unusual barn for the area inasmuch as it is entirely made of stone except for the gable portion. There are entrance doors at both ends and a poured concrete silo at the rear. Two smaller stone sheds were built out from one end of the barn, but they are now gone, according to Richard B. Trask, town archivist of the Peabody Institute. One of the sheds was for wagons and the other for milk and for tool storage.

This barn, built sometime between 1792 and 1800, survives today, along with five other buildings of Shaker construction, on Jerusalem Road in Tyringham, Massachusetts. This Shaker settlement was founded in 1792 by Joshua, Abel, and William Allen; William Clark; Henry Herrick; and Elijah Fay. They worked fifteen hundred acres of fine farmland and owned more than fifty buildings.

This Great Red Ox Barn rises three stories from a coursed fieldstone foundation, and a stone cattle yard sits on the slope below the barn. A covered way entering from Jerusalem Road allows direct access to the haymow, while a second entry to the threshing-floor level is from a circular path under the covered way. The lowest, stone-walled area contains box stalls for cattle.

The Tyringham Shakers traded their Tyringham lands to Dr. Joseph Jones for land in Pennsylvania in 1874. This barn is now privately owned.

Phil
Ziegler
83

The Vigeland Barn, *circa* 1883, is located on Academy Hill in Conway. It was built by John Packer, a local farmer and banker who also served as a member of the state legislature.

This huge two-story barn measures seventy-five feet by forty-five feet. It is constructed of pine and chestnut and topped with a slate roof. The foundation is entirely of stone, there is an internal silo, and it is topped with an old, broken-down cupola. Instead of regular stairs, it has slanted risers, which made it easier for cattle to maneuver inside the barn. Doors lead directly from the barn to the house on both floors. This drawing shows exactly how the barn looks today.

The Earle H. Streeter Barn (1794) is really a sight to behold both inside and out. All the beams are hand hewn and held together with wooden treenails — not a nail in sight. Mr. Streeter, the present owner, describes his barn with pride: ''It is an old-fashioned basement cow barn with a million posts under it, and if people want to see something from Heaven, they should take a look at my 'angel.' '' He also notes that all of the boards in the barn were sawn by an up-and-down saw and that his barn didn't lose even one shingle in the hurricane of 1938.

Located in Cummington, it is indisputably the oldest barn in the area and was in full use until October 1959, when Mr. Streeter's huge, modern barn was completed.

Mr. Streeter housed his cows in the unusually deep, open ''basement'' under the barn proper. This is an uncommon arrangement; in regular cow barns the animals are kept on the ground floor, which is built directly on the stone foundation.

Phil
Ziegler
82

Barns of New Hampshire

The Tomapo Barn, once located in Lebanon, was owned by Howard Townsend and his son Bruce until it deteriorated to the point where it had to be demolished and the parts sold. It is now the major part of the framework of a house in Union Village.

Its historical background is somewhat vague, as the barn was moved to Lebanon in 1850 by Howard Townsend's great-great grandfather. No amount of research could ferret out the date or place of its original construction. It was used as a dairy barn and remodeled three times after 1900, but previous to that the Townsends had run a well-known sheep farm. This farm was also honored with a listing as a bicentennial farm, meaning that it was at least two hundred years old in 1976. The Tomapo farm is one of fifty-seven New Hampshire farms selected for that designation by the State Historical Society.

The unusual Indian-sounding name is the creation of Howard Townsend, who explains that he chose the name mostly because he likes the sound of it and it begins with the first two letters of his last name.

The Towle Barn, built in the late 1700s, was selected as a bicentennial barn during the national celebration of that event. It has been in the Towle family for eight generations and is located in Hampton. Although the house shown in the background of this drawing was built in 1876, the rest of the buildings on the farm are the original ones except the well house, which was constructed in 1916.

Thirty-five by seventy feet, this large barn rests on sixteen granite posts. All of the original framing — hand-hewn beams and up-and-down-sawn boards — remains intact in spite of the age of the barn. The plates and sills are all of one piece, the longest extending the full seventy-foot length of the barn. The flooring is finally giving way, so this barn is not now in active use. It had been used mainly for hay storage, filled with hay from floor to ridgepole by late summer.

Phil Ziegler '83

The Shirley Barn, *circa* 1860, is located in Goffstown, on the Shirley Hill Road. It is still in the family, now owned by Mr. and Mrs. Lawrence W. Shirley. The barn was built by Robert Moore Shirley, who two years after building it, deeded it over to his son Edward Carlton and moved to Manchester, where he died in 1889. The first Shirley to settle on the land, in 1761, was Thomas Shirley, and it has been in the family ever since. Each generation has been very active in local, state, and national government.

At some time, for no apparent reason, the cupola was removed. Subsequent generations have made other changes in the barn, but on the whole, the interior is the same as when it was first built. This drawing of the 120-year-old barn as it looks today shows a perfect example of what happens when loving care is continued.

Originally built by Patrick Connary in the late 1700s, this farm has been in the Shores family since 1825; only Henry lives there now.

Shortly after the farm complex was built in Lancaster, New Hampshire, the Catholic Diocese decided to build a church in nearby Northumberland. They planned to use the Shores house as a meeting place until the church was finished. The priest called for the first meeting, and when the appointed Sunday came he only found one person there, stoking up the fire. No one showed up, and the man who was keeping up the fire remarked that it was a Lost Nation out there. The name stuck, and the road on which the farm sits has since been called Lost Nation Road.

This barn has an extremely rare feature. As shown in a previous chapter, a plate usually runs the length of a barn, tying the bents together. In this barn there are *three* plates running the length of the barn on the windward side (the building face shown in the drawing). Mr. Shores explained that records show that high winds had plagued the area, so the original builder installed the three plates and tied extra-large rafters into the ridge beam in order to prevent the wind from pulling off the roof. It worked; this roof has withstood winds up to 130 miles per hour.

The house also exhibits similar attention to detail. The hand-hewn beams in the kitchen were hand-rubbed and varnished several times by Mr. Stores's grandfather, giving them a beautiful patina. All the buildings look as straight and true as they must have the day they were finished.

Phil
Ziegler
83

In 1722 King George granted the land upon which Nottingham, New Hampshire, was founded. In 1728 Andrew Simpson, an ancestor of the present owner, Mrs. Smith, purchased ten acres of that land upon which to establish his homestead. Eventually the acreage was increased to 150 acres. Andrew Simpson's wife and two male neighbors were massacred by Indians in 1747.

The legend "Built in 1826" was found written in back of one of the horse stalls and also on a supporting post of the carriage shed, thus ascertaining the year the Eben H. Smith Barn was raised. It measures seventy-two feet by thirty-seven feet, nine inches, and is of hand-hewn post and beam construction, with beams eight inches square. The plates are of one piece as are the sills, and the flooring consists of boards ranging from eight to twelve inches wide. These boards are worn in deep grooves, and the wooden pegs securing them look like they are ready to pop out along with the hard knots — but of course they won't. The original eighteen-inch iron hinges are still on the large barn doors.

Oxen, cattle, sheep, and horses have occupied the barn, and the original shingle roof has been replaced with a metal roof.

Although parts of the Frye farm complex date back to 1740, the main barn illustrated here was constructed in 1890. Huge by any standards, this barn is one hundred fifty feet long and fifty feet wide. It is located in Wilton.

Split granite slabs, eight to ten inches wide by eighteen inches deep, make up the foundation under the entire barn. Hand-hewn white oak and chestnut were used for all the beams, which are mortised and tenoned and wooden-pegged. Hand-hewn posts twelve to eighteen inches square support the ground floor of the entire barn. All windows are framed on the outside with molding (a very unusual feature in old barns) and matched tongue-and-groove hemlock covers the sides of the barn and all the outbuildings. The barn's two cupolas are topped by decorative weathervanes in the shapes of animals: a cow on the left, a horse on the right.

Phil
Ziegler
83

Built in 1854, this barn is but one of many sturdy structures in this old New Hampshire Shaker community known as Lasalette of Enfield, or Chosen Vale Shaker Village. Most of the twelve major buildings in the complex are of stone.

The first illustration shows the cow barn from the front and the stone-supported ramp leading to the third level.

The second is of the barn's rear entrance, not accessible to any vehicles, but featuring some magnificent stonework on the facade; this is depicted in the third drawing.

Barns of Maine

When I started compiling information for this chapter I was advised to seek the help of Mr. Frank Beard at the Maine Historic Preservation Commission. When I arrived in Augusta to meet with him, he started right in, sharing his considerable knowledge of Maine's historic barns. He regaled me with truly fascinating stories about the various barns, and with each story he pulled out a slide or photograph from his well-documented files. Never had I seen such an array of beautifully preserved barns. The complex grillwork, the fine stone masonry, the carvings and filigree work were there in abundance. Some of the best examples are presented in the following pages.

The Wedding Cake farm complex, *circa* 1826, was built in the late Federal style for George W. Bourne, whose shipyards were on the Kennebunk River not far from the house. The barn and its attached wing were in the Gothic style as opposed to the Federal style of the house. One day, while standing across the road looking at his homestead, Mr. Bourne suddenly realized just how architecturally monstrous the combination was. He immediately decided to hide the house under a frosting of buttresses, arches, and "gingerbread" so it would match the barn and wing, which already had their share of these embellishments. Wanting the new carvings to closely match those on the barn and wing, Bourne carved them personally with the aid of a young apprentice ship's carpenter, Thomas Durrell.

With typical backwoods humor, Kennebunk residents began to tell visiting tourists about the young, newly-wedded sea captain who spent his time at sea carving objects for the house, installing them one by one when he returned home.

Mr. Bourne died in 1856, but fifty years passed before his home was called The Wedding Cake House. It seems that in 1906 one Frank Bonser conceived the idea of creating a series of postcards based on the barn's embellishments; on these, he referred to the barn as The Wedding Cake Barn, and it has been known as that ever since.

The Gage-Rice Estate is of Federal architecture. The house, *circa* 1817, is of frame construction with a clapboard exterior. It has several attached buildings, but the barn, built about 1825, is separate from the rest of the complex. This farm is an integral part of the Waterford Historic District, which comprises a cohesive and compatible grouping of eighteenth-, nineteenth-, and twentieth-century buildings. It is located on the western shore of Keoka Lake, formerly called Thomas Pond, in the central part of Waterford Township, and occupies an area known locally as "The Flat" owing to its level configuration in the midst of a hilly region at the foot of Mount Tire'em.

As in a typical rural village in Maine, the houses and other buildings are, for the most part, generously spaced. Within the last hundred years or more, Waterford has experienced very little change and appears today very much as it did in the mid-nineteenth century. All of the structures in this historic district are in good-to-excellent condition.

Chadbury-Stanley Farm is the oldest one in the Harrison area. It was purchased by Revolutionary War veteran Moses Whitney in 1806, just one year after Harrison was chartered as a town. The barn was built in 1810, and the farm subsequently changed hands several times before Arthur Williams and Mary Thompson Carlson purchased it in 1955. They still own the one-hundred-acre property, which is about the only farm in the area that has not been broken up into smaller parcels. Electricity was installed in 1955 but there is still no running water.

This barn is built of hand-hewn beams with wooden treenails sealing the joints. It has not been appreciably changed in all the passing years except for the addition of a few horse stalls and two fairly new six-by-twelve-foot barn doors that Mrs. Carlson made herself after the original ones finally rotted away. The "new" doors, however, still have the three-foot iron hinges that graced the old doors.

The Laudholm Farm is truly a slice of Maine as it existed before the hustle and bustle of tourism took over the land that surrounds it. The century-old farm has a great expanse of farm fields and many stands of beautiful trees bordering on a salt marsh and sand-spit leading to the open Atlantic.

Located in Wells, and owned, until recently, by one family, this 250-acre farm is about the only undeveloped saltwater farm remaining on Maine's southern coast, but if it has its way, a concerned group of Wells citizens will change all that and a true "working farm" museum will be created. All the group needs to accomplish this goal is $750,000.

By happenstance, this acreage overlaps the northern limits of the range of the cottontail rabbit and the southern limits of the snowshoe hare's range. Harbor seals, mink, otters, muskrats, and raccoons are a common sight, along with ducks, herons, and even an egret or two. An occasional bald eagle has been seen although no nests have been found.

The Norumbega Carriage House, in Camden, is of fieldstone and frame construction with an L-shaped plan, cross-gabled roofs, shingle siding (except where fieldstone is exposed), and a tall, external tower attached to the east corner of the building. This tower is cylindrical, with a conical roof; the lower half up to the cornice of the carriage house is stone, and above that it is frame, with shingle siding. Just below the bracketed cornice of the tower, several windows of varying shapes are positioned.

The facade of the carriage house faces northwest and features an arched barn doorway in the fieldstone section. Above this is a second-story arched bay with double doors flanked by small circular windows. There are two other ground-level entrances: a hooded one to the extreme left end of the facade and another to the right of the carriage entrance. An extension of the building to the southwest, which creates the L-shaped plan, contains a second barn door facing northeast. Above this carriage entrance is a double doorway with two 28-paned windows.

This complex was built in, or shortly after, 1886.

The Clock Farm of Kennebunkport, Maine, built from the mid-nineteenth century on, is a remarkable grouping of connected buildings, the most distinctive of which, the clock barn, lends its name to the entire complex.

The attached 1½-story barn has a louvered square cupola atop its gabled roof. In front of that, astride the ridge of the barn, is a tall, square clock tower. Each side of this tower contains a large four-faced clock with roman numerals below a segmented, hipped roof fitted with semicircular louvered vents.

Somewhere in this collection of Greek Revival buildings (probably the ell immediately to the left of the main house) is the original dwelling built on this site by Peter Johnson in 1773. However, the most important feature of this landmark is, of course, the clock tower.

At some time in the late nineteenth century the farm was purchased as a summer residence by Thomas Emmons, a Lawrence, Massachusetts, manufacturer. The clock had formerly been at the Emmons Loom Harness Company, but the clock reportedly kept such bad time that it caused some disagreements between Mr. Emmons and his workers. Whatever the reason, Mr. Emmons had his clock transported to his summer residence, where, presumably, knowing the correct time was not so important.

The Nutting Homestead, built between 1796 and 1820, has been occupied continuously by the Nutting family for over five generations.

Nathan Nutting, born in 1761 in Groton, Massachusetts, moved to Otisfield with his family in 1796 and built the original Cape Cod section of the house. His son, Nathan, Jr., born in 1804, was, according to family tradition, sent to Boston as a boy to study architecture and the building trade. Nathan in 1820 added the attached Federal-style house to the original structure.

To the north of the homestead stand three nineteenth-century agricultural buildings: a carriage house, barn, and corn crib. Like the frame house, these buildings are basically in their original condition.

Although Maine has long been an agricultural state, it is rare to find an authentic nineteenth-century farm complex such as Nathan Nutting's. The present owner, Albert D. Nutting, continues his family tradition of carefully maintaining the homestead.

The Sam Perley Farm, near Naples, is a simple but distinctive example of Federal architecture. It was completed in 1809, and any subsequent additions or repairs have been minor.

The large, gable-roofed barn is mostly shingled, with the bottom half of each of the long sides finished with vertical barn boards. The entire foundation is supported by a series of stone columns. The square, ventilated cupola gracing the roof, with its covered top ornamented by a lathe-turned pole at its peak, is shown also in close-up; a great example of this type pf cupola.

The boarded-up structure to the right of the barn is an old well house. Although not in the true sense a gazebo, it may at one time have been used as such.

The Sam Perley Farm is named after the builder's son rather than after the builder himself. It is situated on a hill in a remote part of Naples Township. It is significant both for its architectural merits, and because it remains virtually unchanged as an early nineteenth-century farm. Enoch Perley, who built the barn, came to the area as one of the first settlers in 1776; he had been a member of the Bedford Company of Minutemen who marched to the defense of Concord on April 18 and 19, 1775.

Phil Ziegler '83

Built about 1867, the barn at the Ebenezer Alden farm in Union is a very impressive structure measuring forty by eighty feet, with hand-hewn posts and beams throughout and joints secured with wooden pegs. Its four stories give it considerable height and it is completely shingled on all sides. Its interior is impressive enough to have inspired members of the staff of the State Museum to call it the "Cathedral of the North," and the main house, built in 1797, is listed in *Maine's Historic Places* (see Bibliography).

Of particular interest is the ten-foot-long carved eagle installed over the main entrance to the barn. It was commissioned by the husband of Mrs. Hazel Marcus, the present owner of the barn, and is a replica of the eagle found on the sternboard of a sailing vessel built in the shipyards of Dunn and Elliot of Thomaston, Maine. Before the present eagle was installed, the space had been occupied by one of four cast-iron eagles reputed to have been made by Paul Revere to adorn "Montpelier," the Thomaston mansion of Revere's friend, General Henry Knox.

When the original "Montpelier" was torn down in 1871 to make room for a railway, its contents were sold at auction. Ebenezer Alden's son, Horatio, bought the wrought-iron gate and iron fencing, which he melted down in his foundry to make anchor chains. He also purchased one of the cast-iron eagles and hung it on the barn, where it remained until 1965, when it was sold to an antiques dealer.

This carriage house, constructed by George Curtis in Bowdoinham, *circa* 1875, is a highly ornate example of domestic architecture in the Italianate style. The Cornish Carriage House is two stories high, with a gabled roof and clapboard siding. The cornices are elaborately bracketed, and the barn door and windows boast miniature carved barge boards at each corner. An ell connects this structure with the main building, and both house and ell carry the same decorative detail as the carriage house. The Bowdoin *Advertiser* of September 11, 1885, commented that this was one of the best residences in town and an ornament to the street. The present owner is in the process of restoring this lovely complex.

Phil
Ziegler
83

The Northeast border of Canada and the United States marks the junction of Acadian, Quebecois, Loyalist, Yankee, and many European immigrant cultures. Between Allagash and Houlton, Maine, and St. Francis-de-Madawaska and Woodstock, New Brunswick, the main industry is agriculture. The barns, many over 100 years old, sustain the alliance of these people to agriculture, particularly to the raising of potatoes.

Foremost of these barns is the semisubterranean "potato house," whose stable temperature insures cool storage space in the summer and protection from frost in the winter. The variety of such barns seems to be limitless.

The Canadian barns generally fall into one of three categories: (1) the Acadian, a simple side-entry, gable-roofed barn with one main central threshing floor; (2) the Quebec barn, basically the same as the Acadian, differs in that there is a series of threshing areas that open to several side entrances; (3) the Dutch barn is usually characterized by its large size and a gambrel roof. It is mainly used for dairy purposes and is found near highly populated areas. Characteristic of the Dutch barns are unusually large threshing floors, a substantial haymow, and livestock housing on the ground floor or sometimes in a basement area built especially for that purpose.

For reasons unknown to the author, the Madawaska twin barn style illustrated here is restricted to the settlement-expansion areas of Maine's upper St. John Valley. Although adapted for use on the Maine side of the St. John Valley, the Madawaska twin barn did not originate there; identical structures are found in several Quebec communities between Sorel and Drummondville, south of the St. Lawrence. Most of the twin barns in existence today date from the 1920s, which was a time of tremendous expansion of farming in northern Maine.

Twin barns, like this one in Madawaska, consist of two identical structures placed side by side with the long sides parallel, connected by a passageway varying from one story in some examples to roof-ridge height in others. Usually, one barn held the livestock stalls, along with a haymow and tool-storage area, while the other half included the granary and threshing floor as well as storage space for the farm machinery and perhaps another haymow. The passageway often contained pens for hogs and poultry, which always had access to a fenced-in portion of the barnyard.

There are many cases where two barns of similar dimensions were moved from different sites, reassembled side by side, and connected by the usual passageway, thus completing the unit.

The earliest twin barns had gable roofs, while those built after the 1920s usually had gambrel roofs with connecting ridges.

The twin barn is indigenous to the American side of the St. John. It took care of the demand for enclosed, weather-tight space as well as the extra storage needed on the American potato farm. Potato-cultivation equipment was very expensive, and you can be sure that it was well taken care of by the owners; the large twin barn was ideal for this purpose.

Phi Ziegler 83

Phil
Ziegler
83

Another major type of barn found on the American side of the St. John Valley is the Aroostook barn, a northern version of the New England connected barn. In Aroostook, general purpose barns for storage and livestock are usually connected to each other when several structures are located on the farmstead. These barns are rarely identical in size or shape, and they are invariably connected in tandem or at right angles to each other. They appear in both the side-entry or gable-end versions and are usually clapboarded and painted white. Interior arrangements vary considerably from barn to barn to meet the individual farmer's needs.

The huge Aroostook barn illustrated here was designed and introduced to the area by New Englanders of Dutch descent who migrated to the St. John Valley. It is located in Grand Isle.

One such structure, known as the Hoyt-Wheeler Barn, was 87 feet by 175 feet and took over 325,000 board feet of lumber to build. It could accommodate more than fifty cattle and twenty-five horses under the haymows and an additional seventy-five cattle under the main floor. That magnificent example of an Aroostook barn burned to the ground in 1919.

Phil
Ziegler
83

Barns of New York

Because many wealthy Dutch immigrants acquired large grants of land in this state, large barns were a necessity. They were called Plantation barns.

Steven Van Rensselaer was one of the most favored immigrants. His grants of land were likened to the plantations of the South and his house was referred to as a manor. Procuring labor to work these tracts of land became a serious problem for such grantees, and, although it is not now generally discussed, slavery became prevalent throughout this section of the North, even as far up as Albany. Slavery continued in the North as late as 1840, when public attitudes turned against it. The big manor owners were finally forced to dispose of their slaves. This led to the break-up of the big plantations into smaller parcels that were either purchased or leased by less affluent farmers. Those who leased land soon raised enough money to purchase it, and then there were very few leased farms left.

Wheat was the major crop of the plantations, but with the breaking up of the manors it was no longer a profitable one, and soon the smaller farms shifted to dairying.

The Altamont Raised Dutch Plantation Barn, built about 1750, was originally part of the great Steven Van Rensselaer manor. With the changes in the agricultural scene around 1840 came the necessity of changing the barns into dairy barns. Space for the animals and a second floor to store the hay needed for them had to be provided. This was done by literally "raising the roofs" of the existing barns.

First, the roof was carefully removed to make room for a second floor. This alteration was made by creating a series of new and smaller bents on the top of each existing cross-beam, using eight-foot-long vertical beams. These bents were then connected at the top, lengthwise, exactly as the original plate had been installed. This long beam was called a purlin beam, and those at the top of the new bents were called purlin crossbeams. Notches were then cut into the purlin to match the rafter placement of the original roof so the rafters could then be set back down in their original positions on the plate when the roof was put back on. This created a strong new roof and added much-needed room for hay storage. The Altamont Raised Dutch Plantation Barn, illustrated here, underwent just such an alteration.

For a detailed drawing of the above procedure, refer to Chapter 2, "Construction and Barn Raising."

This is called the Date Stone Barn because a stone was found in one of the horse stalls with ''built in 1875'' chiseled on it. ''Raised June 24,'' presumed to be the exact day the original barn was finished, is carved in the wood of the same stall.

Located originally at Fort Orange, which is now Albany, the barn was carefully dismantled and reassembled at Altamont, where it now stands in nearly its original condition — but not its original shape. Doubtless this same thing happened to many barns that were originally located along the rivers. When developing industry drove out agriculture, the barns had to be dismantled, removed to remote areas, and reassembled. Although all of the original parts were used in the reconstruction of these barns, in most cases their shapes were modified as each farmer rebuilt his barn to meet his individual needs. Thus, old barns became new barns, and when the date was recorded it was usually the date of the *reconstructed* barn, not the date of original construction.

The present owner of the Date Stone Barn is Ms. Beverly Waite, who is proud of her ''new old'' barn.

This large, gambrel-roofed barn on the Eastern Defreest Homestead in North Greenbush was built about 1900 for the Jordan family dairy farm. It is located east of the house and replaces the original barn, which had been situated to the west. The barn is approximately eighty feet long (east to west) and forty feet wide (north to south). The ground floor has doors on the east, west, and south sides; the north side is below grade. There are a bullpen and two stalls still in place on the western end, but the other cow stalls have been removed. The upper floor (hayloft) has an entrance on the north side. The barn contains one small room that was probably used for feed storage, as it is very tightly constructed. A small milkhouse stands outside the west door, as illustrated, and the foundations of a long-gone silo can be seen on the south side.

The Defreest Western Homestead Barn, located at North Greenbush, resembles in basic proportions and construction a New World Dutch barn. However, there are differences that suggest it is not an early example, but instead a nineteenth-century (transitional) barn type. The barn is nearly square, measuring fifty feet by forty-five feet. (A later addition on the south side of this structure is not illustrated here.) The gable ends of the original barn are oriented east and west, with the wagon door on the north side. The livestock area on the ground floor has an entrance on the east side.

This barn has the typical Dutch H-frame structure and a center-aisle threshing floor. However, although the New World Dutch barns had the threshing floor on the ground level, this barn does not. Also, the threshing floor does not run from gable end to gable end, as is expected in a New World Dutch barn. Another difference is the size of the anchor beams and the tenons, cut flush with the mortises in the columns of the Defreest barn.

The differences evident in this barn show Dutch influence but the bi-level construction indicates that the barn was built during the early nineteenth century. The presence of many eighteenth-century characteristics suggests that the framing of an earlier barn was reused in the construction of this barn. As the site was occupied at least by 1740, the earlier barn probably existed nearby.

According to the owner, Mrs. Mildred Halprin, of Malone, this barn was built over two hundred years ago. Its pristine condition suggests much loving care from generation to generation, and all of the owners of this barn from its builder to its present owner could well take pride in their contribution to preserving a beautiful example of our agricultural heritage.

The Halprin Barn is built into a hill on three levels and is supported by a sturdy stone foundation. The hand-hewn beams and columns were beautifully made and of good fit, for the barn seems to be still as straight and true as the day it was built.

Phil Ziegler
83

Not many barns are found in the Adirondacks in the Indian Lake area. The Camp family barn is one of the few existing examples. It was built about 1891, according to Ted Aber, historian for Hamilton County. Mr. Aber exlains, "This barn is not really unusual in that, although large, it is merely an oblong barn without any frills."

This barn, *circa* 1746, was located near Knocks, New York, before it was dismantled and transported to Framingham, Massachusetts. Little seems to be known of the history of this German plantation barn, but this drawing shows it as it stood on its original site.

Located near Guilderland, this eighteenth-century Dutch barn is purported to have been built *circa* 1750; it certainly does have the general appearance of the barns built during that period. Very few historical facts about this barn are available, and sad to say, it has finally collapsed. All that is left is a pile of rubble.

Phil
Ziegler
85

The Dutch Colonial homestead known as Mount Gulian was built between 1730 and 1740 by Gulian Verplanck on land bought by his grandfather from the Wappinger Indians in 1683. This homestead at Fishkill is of great historical interest: It served as the headquarters of General Von Steuben during the final period of the Revolutionary War. It was also here that the Society of Cincinnati came into being in 1783.

Adjacent to the house is this restored eighteenth-century barn originally owned by Philip Verplanck. It is unique for a New World Dutch barn because of its overhanging gable ends.

The term *mount* was often used in the names of country homesteads whether or not they actually stood on high ground, but in the case of the Mount Gulian homestead, it is appropriate, for the buildings are situated on high ground that gradually slopes down to the Hudson River.

The drawing shows the barn after restoration. Still to be restored are the extensive gardens, which were laid out in 1804.

This eighteenth-century Dutch barn, before being dismantled during the building and restoring of other projects, was lòcated on the Art Smith farm in Coxsackie. It was built about 1750.

Many parts of this barn were used in the Music Building at Wolf Trap in Vienna, Virginia, and at the site of the reconstrucion work done at Phillipsburg Manor at Tarrytown, New York. Complete details of these projects can be found in the chapter titled "Restoration."

The Vermont–New York Barn could very well be the oldest barn still standing in America. According to historical records, it was commissioned to be built, along with many churches, chapels, and forts, by French priests in the St. Croix Terrace area around 1540. The barn was built a few miles south of this settlement. When the Indians later destroyed almost the entire place, the barn was spared. The structure has since been carefully dismantled and each piece numbered so that it can be reconstructed at Richard Babcock's museum in Hancock, Massachusetts.

The barn's name derives from a later time. An immigrant from England purchased it while Vermont was still part of New York. Evidently he wanted a barn in Vermont because after the state boundaries were settled in 1777, and the barn, instead of being in Vermont, proved to be located ten miles into New York, he sold the barn and built another just over the border in Vermont.

This barn contains many elements of construction found in twelfth-century churches and bridges, chief among them the kingpost truss supporting the main swing-beam in the barn. Although Andrea Palladio is credited with designing the kingpost truss and many other kinds of trusses, such as those found in covered bridges, these designs were used centuries before he was born in 1518.

The entire barn is constructed of solid oak. No wonder it has stood for over four hundred years.

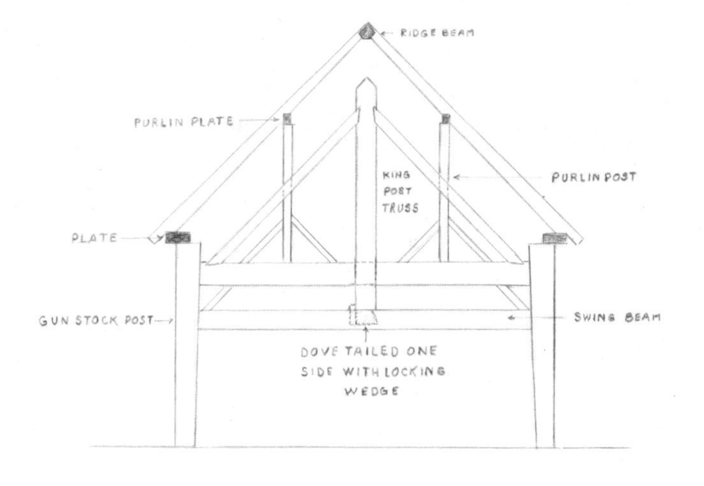

KING POST TRUSS VERMONT-N.Y. BARN

In Schoharie, New York, you will find these connected eighteenth- and nineteenth-century barns. The barn on the right is an overshot barn, built about 1750.

Overshot barns were built with a ramp to the second floor so that hay wagons could drive right into the hay storage loft. The hay was always stored in the second floor area, and the first floor, called the threshing floor, was reserved for the livestock. The term *overshot* is seldom used today; even people who are truly barn aficionados seldom recognize it.

The structure at the extreme left of the illustration is a carriage barn. Such carriage barns, or sheds, were built alongside Dutch barns to accommodate farm wagons and implements; the loft stored excess hay.

These barns are still standing on their original site. Although the farm in general is no longer a working farm, huge amounts of hay are still stored in the barns.

The early-nineteenth-century Messenger Barn is truly a magnificent structure. Sixty by sixty feet square and three stories tall, it is a really huge barn, topped with not one, but two cupolas. One of these is in scale with the rest of the barn, but in front of it is a small cupola the size of a glorified doghouse. The barn has a full basement and is in a sad state of disrepair, but the present owner, after talking with Mr. Babcock, a master barn builder, has decided to restore this barn completely.

Phil Ziegler 82

Barns of Vermont

The original settlers of Vermont, many from Connecticut, had to be a tough, determined group of people. The climate was anything but balmy and the boulder-strewn countryside made farming a risk by any standards. It was a task that discouraged many.

However, those who did stay soon began to work the land, and agriculture flourished. Dairy barns dotted the countryside, for dairy farming was the mainstay and remains so to this day. Some were huge like the Dutch barns of New York and Pennsylvania while others were of intermediate dimensions. On a per capita basis, Vermont remains the nation's most productive dairy state, so it comes as no surprise to learn that most of the existing barns are single-purpose barns devoted to dairying.

The coach barn at Shelburne Farms was constructed in 1901. Its general architecture follows that of the farm barn, which was constructed between 1887 and 1889. Although the farm barn, not illustrated here, is part of a working farm owned by the Webb family and is an absolute marvel of architecture, the coach barn was selected instead as an illustration for this book because it is a perfect jewel of compact, well-planned architecture of the early twentieth century.

The coach barn originally housed coaches, tack, grooms, and horses. It is now owned by a nonprofit organization, Shelburne Farms Resources, and is used only as a base for environmental-education programs. On warm days the barn may be surrounded by groups of young people practicing their musical instruments outside, adding to the beauty of the scene.

The handsome old Ransomvale Farm Barn is a perfect example of what tender loving care can accomplish when it comes to preserving the beauty of a barn. This barn is a 2½-level, gambrel-roofed structure with gable-roofed attached sections and three silos. The gambrel roof was added about 1920.

The silo shown in this drawing is constructed of short staves bound together with steel rods formed into hooplike bands, a style typical of the silos built around 1860.

It can be found in Castleton, Vermont, where it is still in use as a dairy barn.

Phil
Ziegler
'82

Built around 1860, the Armstrong-Howe Barn is located near Tunbridge. It was originally called the Armstrong Place. It is a 1½-story barn with clapboard siding covering the entire building. The foundation is concrete and the roof is now metal. Construction details indicate that the main barn is actually two connected barns.

A newer lean-to is on the east side. In 1950 a round-topped silo and a balloon-framed barn with a sheet-metal roof were added. No reason is known why the large shed next to the main barn was given such a decorative opening.

The beauty of the Pony Barn speaks for itself and shows that it has been carefully maintained since its construction in 1900. Nominated as a historic site, it is located in East Burke and is a perfect example of an estate barn; that is, one designed to complement the main house in terms of architectural style.

Six gables set off the roof, which is topped by an octagonal cupola.

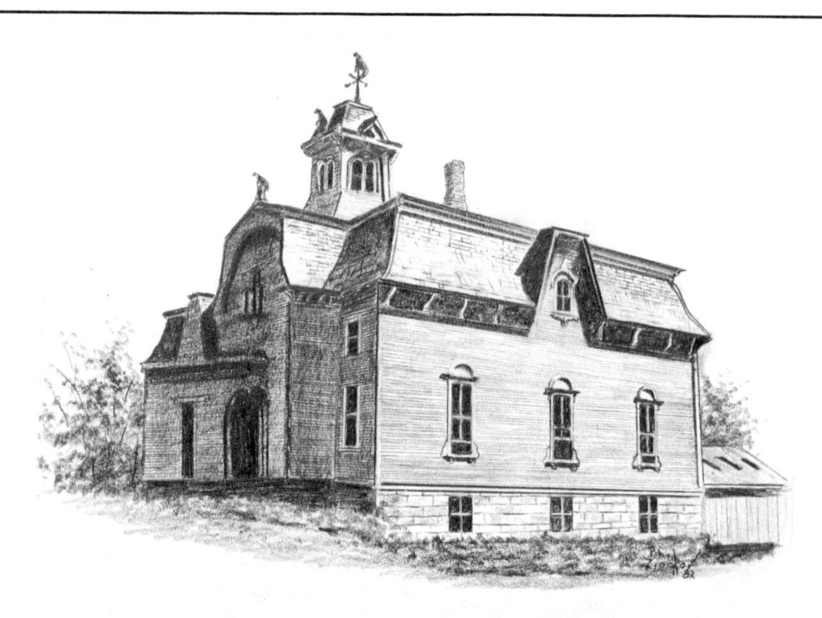

Shortly after 1900 the owner of the Morgan Horse Farm near Weybridge, Colonel Battell, offered the farm to the United States Department of Agriculture. The gift was to include all of the buildings, five hundred aces of land, and all of the breeding stock on the farm. However, Battell stipulated that the farm was to be used only for the breeding and propagation of the Morgan horse. The government initially refused the offer on the grounds that it could not absolutely guarantee to meet that requirement, but a compromise was reached eventually. Although the propagation of the Morgan horse is, in fact, the main activity of the experimental farm, sheep, hogs, poultry, and cattle also are bred there.

The barn illustrated here is a late-nineteenth-century structure, situated near the statue of Justin Morgan by the farm entrance. Built on an asymmetrical plan with a dressed-stone foundation, it has clapboard siding and windows with ornate hood moldings. The mansard roof is slate shielded, and a weathervane with the image of a Morgan horse tops the beautifully ornate cupola.

Owned by Lawrence Rainville, this barn, constructed *circa* 1870, is located near Fairfield, in Franklin County.

The Rainville Barn is of Italianate construction and has a very sturdy stone foundation (not visible in this drawing). It has horizontal clapboard siding, a gabled roof, and post-and-beam construction throughout. This now dilapidated structure was once a very stylish carriage barn, and features a rare cross-gabled cupola.

Although it was sometime prior to 1877, the exact date that the Heath-Mallary Barn was built cannot be ascertained.

The ell attached to the rear of the barn shown in this drawing is now gone. In the lower section the doors and walls have been removed and a large storage space created. Much work has been done on the barn to protect it structurally, although at one time it seems to have been slated for demolition. No changes have been made to alter its general appearance, thanks to the Mallarys. The hip-roofed cupola is intact, and the doors and windows remain as they were when the barn was first built.

This gigantic barn of post-and-beam construction is located in Fairlee.

This gambrel-roofed barn in Fairfield, built in 1923, has been nominated as a historic site on the National Register of Historic Places. Very little information about the barn itself is known, although a complete history of the Allen Geer farm is documented. It is included in this book because of its unusual entryways, the two barn-door openings. The one on the left is for off-loading from wagons, while the right allows them to drive right into the barn via the ramp. It has been nominated despite its relative youth because it is a beautifully constructed early-twentieth-century barn.

The Proctor-Clement Barn, located near Rutland, is a beautiful example of well-preserved, mid-nineteenth-century architecture. This detached, two-story, gable-roofed barn is of post-and-beam construction with horizontal exterior clapboard sheathing. The original roof has been replaced with asphalt shingles, and a single ventilator cupola adorns the center of the roof. Built about 1867, this barn's foundation is entirely of dressed granite, and it would take a powerful earthquake to disturb it.

Phil
Ziegler
82

North of Jericho, this impressively huge barn looms out on the left side of Route 15. Records found in an old box in the attic of the house indicate that it was built around 1832. Originally, the house and barn were used as a way station, and the inside of the house looks much as it did during that period. The barn was the carriage barn; stalls still exist where travelers' horses were stabled while their owners stayed at the house.

The most impressive features of the barn are the huge beams, mostly hand hewn, with the tenons and mortises secured with treenails. Unfortunately, according to present owner David Turner, the barn has suffered some damage because a previous caretaker, unaware of (or unimpressed by) its historic significance, allowed someone to remove the crossbeams with a chainsaw and sell them to decorate a new house. Now the structure echoes like a huge cavern, for all of the upper floor boards have been removed.

The foundation is made entirely of expertly fitted, unmortared fieldstone, and hardly a stone appears to be out of place. A ventilated cupola tops off the barn and two steel ventilators, installed about 1920, still remain on the roof. Inside the barn at the right corner, an immense silo once stood, but nothing remains of it but two blackened walls.where the damp silage stained the wood.

At the very end of a small dirt road stands the old Harley Rogers Barn. Rogers recently sold the old farmstead, but he was glad to tell the story of the barn.

One of his ancestors, Thomas Rogers, moved to the area with his family in 1798 and built a small house on a nearby hill, planning to base his homestead there. They ran out of water, though, and had to move closer to the road, which at that time was nothing but a set of wagon tracks running across a field. They soon commenced building this barn by their new house.

Finishing his story, Harley Rogers fetched in a piece of board that he had cut out of the barn when he left the family farm for good. On it is carved the date 1817. That one board is all he has left of the old family farm.

Phil
Ziegler
83

Phil Ziegler 83

Barns of Connecticut

Connecticut proved to be the most difficult state in which to locate historic barns. Tobacco barns are fairly plentiful, but because of their structure they are relatively short-lived, and the remaining ones are not truly historic. Historic all-purpose barns proved to be quite rare in the state.

Although not a historic structure in the truest sense, this tobacco shed is a fine example of those still in existence. The owner, Robert Silliman, feels that the shed will probably be used for only a few more years. Tobacco acreage in the Windsor area has dropped dramatically in the recent past, and this barn's useful time has just about expired. Built in 1920, the shed stands on the Huntington Farm in Windsor.

Built in 1860, the Ford-Ransom Barn is a very good example of how more space was added when needed. On the left is the old horse barn. The second barn was added to give more space for hay storage. Although the barns are in disrepair, the Ford-Ransom House itself, about a thousand feet away, is in excellent condition. The house and barns are located in Windsor.

Charles Klamkin, author of the book, *Barns,* took the photograph of the venerable old barn that, with his permission, was used as the subject for this drawing.

As Mr. Klamkin recalls, the photograph was taken somewhere between Windsor Locks and Suffield; however, all attempts to identify it further have failed. Therefore, this drawing is simply a tribute to all of these old forgotten barns, each of which played its part in the development of our agricultural heritage.

The old Ward H. Clemons Barn, located near Cheshire, was, according to the Town Clerk, built around 1700. It is a very imposing structure of hand-hewn, post-and-beam construction with all the joints secured by wooden pegs (treenails), and supported by a stone foundation. At some period, a three-bay carriage house was attached to the barn, housing vehicles as well as grooms. Also, a portion of the barn was once an active blacksmith shop, and many relics relating to blacksmithing have been found in it. Eventually, only hay was stored in the barn until even that became too costly. Today it stands as nothing more than a treasured relic.

Phil Ziegler '83

The Platt Barn is a handsome old double barn, built in 1879. It is located at the junction of Route 172 (South Britain Road) and Flagg Swamp Road in Southbury. Edwin Pierce acquired the property on which this barn was built in 1869, and after a succession of different owners, Sidney Platt and his wife purchased it in 1921.

The saltbox roof on the barn in the foreground gives it a sweeping symmetry with the attached barn in the background. The cupola and the delicate carvings below the arched windows must have made it a beautiful sight in its heyday.

The Hazardville Powder Mill Barn in Enfield was built about 1845. Though used as a regular barn at first, it was converted to a powder-manufacturing facility during the Civil War, shipping much-needed gunpowder to the Union army. Its basic structure and proximity to shipping facilities no doubt prompted the owner to convert the barn to that hazardous use.

Hand-hewn post-and-beam construction secured by treenails is evident throughout the structure. The plates and sills are eight inches by ten inches and over forty feet long. Although wooden-pegged mortises and tenons are predominant throughout, some steel bracing joined with nuts and bolts was used, possibly in those places where heavy machinery and lifting devices were installed during the gunpowder-making days.

Living quarters for the present owner, Mr. Ralph Sweet, are on the second floor, and what was once the threshing floor is now a dance hall and auction barn. It is so sturdily built that not even an inch of sagging is discernible along the roof line.

This eighteenth-century barn stood in Middlefield before being dismantled sometime in the mid-twentieth century. It was a barn of considerable proportions. The second drawing shows some interesting features in it as well as one from another barn.

Figure 1 shows a plate and corner post that has "through" tenons secured with treenails; the plate is recessed to receive the post.

Figure 2 shows a through tenon secured by treenails.

Figure 3 is a good example of how parts were roughly numbered by the builder so they would match up when assembled. Each brace has a number that is matched with a like number on the post or beam where it is secured.

Figure 4 is an illustration of a canted purlin and queen post, quite common during the period (*circa* 1835) when this barn was built. This canting of the purlin and its supports permits it to meet the rafters at a ninety-degree angle; in earlier frames, the purlin was placed vertically and had to be notched to receive the rafters.

Phil
Ziegler
83

Connected Barns

The basic concept of the connected barn existed in Europe for many centuries and is also prevalent throughout Canada. However, there are significant differences between the Canadian and American groupings. In the American version, structures were added one by one as needed, forming a pattern of irregularity in the roof line, while in the Canadian connected barn the roof line is practically straight, much like a row of freight cars placed end to end. In some instances, these random connections evolved in the form of a square — much like a frontier fort, except that the purpose of the design was convenience, not protection.

The American version is most common in the northeastern United States and so is referred to as the New England connected barn. The shrewd New England farmer decided that it was foolish to brave the elements when he had to do his chores if another solution was possible. Already having a row of larger structures, he knocked a hole in the house wall nearest the first outbuilding and put in a doorway. He then constructed a smaller ell connecting the house to the first barn building. This small shed in many cases was used for wood storage. This process was continued until a row of buildings, all connected by smaller sheds, ended with the main barn. Perhaps one would be a chicken coop and the next a sheep pen or a milking shed, and so on. Most of these animal sheds opened into fenced-in segments of the barnyard, although the fowl had the run of the place.

While this arrangement was convenient, it was also hazardous. All of these buildings were of wood construction, and fire was a constant enemy. When a fire did get started, it usually meant the total destruction of the entire complex. To combat this, some farmers removed the sheds next to the barns and those connected with the houses and sealed up the door to the house again. In one small village in the early 1600s, the town fathers posted a proclamation that if any farmer built a barn or haystack within six "polles" of a dwelling, a fine of twenty shillings would be assessed against the owner. Similar laws were passed by other New England villages, but they were so unenforceable that the law was soon forgotten. However, while many farmers did keep these complexes intact, they kept a close watch over them.

Several different examples of connected barns are illustrated on the following pages.

This drawing of a connected barn of log construction is based on a photograph found in Eric Arthur's book, *Barns,* and used with his permission.

Although Mr. Arthur indicates that this barn was located in West Danville, Vermont, no concrete evidence as to the exact location or history of the barn could be found — hence I call it the Mystery Connected Barn. However, it is an excellent example of the earliest type of connected architecture — a series of log sheds and barns connected to a stone house.

This connected house and barn were built by General Lewis Morris around 1775 in the town of Springfield, Vermont. The house is based on the Georgian plan: two rooms flanking either side of a large hall, with a chimney stack on either side of the hall. The attached barn and carriage shed are in excellent condition and have been beautifully preserved. The buildings are now sales rooms for antiques.

This farm, constructed between 1724 and 1760, is now owned by the Society for the Preservation of New England Antiquities and is open to the public. Over three hundred acres of open fields surround the beautifully preserved complex. The buildings depicted in this drawing are near the central-chimney, two-story house built by David Cogshell of Newport, Rhode Island. A very early one-room ell projects from the east end of the house.

The farm is of historical, as well as agricultural, importance. Cogshell's son-in-law, Silas Casey, was an East Greenwich merchant who lost four sailing vessels during the Revolutionary War. The exterior of the house is scarred by bullet holes, the result of a skirmish between local patriots and British sailors. Civil War general Thomas Lincoln Casey, who regularly summered at the Silas Casey farm, was Chief Engineer of the United States Army and supervised the construction of the Washington Monument and the Library of Congress.

The farm is located in North Kingston, Rhode Island, and was given to the Society for the Preservation of New England Antiquities in 1940 by Edward Pierce Casey.

This grouping has the uneven roof line so typical of New England connected barns. All the buildings stand straight and true, including the silo, which has a wood shingled elevator on one side.

This connected barn grouping (*circa* 1907) on the A. Anthony Farm is a perfect example of excellent planning. Note the practical distribution of all of the connected buildings.

The two main structures, the barns, are neatly kept both inside and out. The gambrel roofs add a pleasing symmetry, and a stone wall set off with hedges surrounds the farm buildings. Located in Middletown, Rhode Island, this connected barn is now a commercial nursery.

Built about 1850, the Albro Farm is a fine example of Federal architecture. While the New England tendency to connect one building to the next is present, there is no evidence that the entire complex, including the house with its large brick center chimney, was connected. The siding on the barn and all of the outbuildings is wooden shingles, as are the gabled roofs. Although clapboard was the most common material used during the mid-nineteenth century, shingling is predominant on this farm. Why this is so is not clear. There is no evidence that the shingles were applied over an older clapboard surface.

One can find this connected barn on Mitchell Lane near Portsmouth, Rhode Island.

Built in Harpswell, Maine, between 1830 and 1897, this well-preserved connected barn property (also referred to as extended architecture) consists of a linear series of attached frame buildings with gable roofs and clapboard siding resting on granite foundations. The main barn was constructed in 1894. It has two stories with a beautiful ventilated cupola topping the gable roof.

The principal entrance to this farm faces the south. The buildings are, starting from the right: the barn, two ells, the main house, another ell, and finally, an older barn (not illustrated here), which was built in 1757.

Thomas Skolfield, son of a wealthy Irish landholder, came to America as a young boy in 1757. He was graduated from Oxford University and for a time taught at the Boston Latin School and at Harvard College. Upon his death in 1796, only one of his five sons, Clement, remained on the land and vigorously worked the farm. Clement had seven sons, and of those, only George, known as Master George, stayed on the farm. While he ran a very successful shipbuilding business across the road he also preserved this lovely farm, as the family does to this day.

Circular Barns

In this chapter, the term *circular barn* encompasses many different types of buildings: There are dome, octagonal, hexagonal, and polygonal barns, among others.

Many theories have been presented as to just how this shape of barn came into existence. One is that farmers with a flair for mathematics realized that a circle as a geometric figure encloses the most area for the least amount of wall space, thus reducing construction costs. Also, the circular barn was found more efficient in terms of distribution of stall area for the livestock and simplified the chore of cleaning out these stalls. In addition, the design provided a more spacious threshing floor. Other builders of circular barns were religious zealots; rather than attribute the design of these barns to more practical reasons, they claimed that they built them to eliminate corners for "the Devil to hide in."

Thomas Jefferson was so greatly influenced by the trend toward circular barns that in 1806, when he adapted this design to a summer estate he had built for himself in Bedford County, Virginia, he had even the privies (called "gazebos") constructed in octagonal shapes.

President Madison and his wife lived in The Octagon, an eight-sided residence built about 1782, for several months in 1814-15. They had fled the White House in August 1814 when British troops stormed the city and burned the White House and other public buildings.

Although the truly round barn was the first of such designs to appear upon the agricultural scene, it was replaced by the polygonal barn when barn designers began to realize that this was a more practical form.

A round barn had to be particularly large because the three-to-four-inch-wide clapboards (the most favored width) produced at the time tended to be very short and therefore hard to bend to fit a strictly round barn. Another drawback was that once a round barn was finished it was almost impossible to add to it, should more space be needed. In the polygonal barn, both of these problems were eliminated while the pleasing circular appearance was preserved.

Unfortunately, like most of our old barns, the circular barns have also become victims of progress. They are disappearing, one by one, until the day may come when even to see one will be a treat. Probably the most notable of all existing circular barns today is the famous Shaker Round Stone Barn in Hancock, Massachusetts, which, with its central silo and fascinating stone walls, is both representative of the circular design and beautiful in its own right.

The Shrewsbury Round Barn, most beautiful of the round barns, is in Winslow, Maine. Although not built until 1913, it is included in this book because of its unusually clean lines and because it, like the majority of barns depicted, has been nominated as a registered historic site. Unique in every way, it resembles a huge planetarium when seen from the air.

The barn is fifty-six feet in diameter and forty-two feet in height. It's entrance is graced by a large frontal projection with an arched roof. The footings, ground floor, and silo are of concrete construction, while the walls are wood-framed. The most distinctive element of the barn is the centrally positioned silo, which is fourteen feet in diameter. This location makes the distribution of silage a simple task.

This barn, so unusual in its architecture and general design, has received much attention in architectural and agricultural journals. Its final construction was the work of James N. Dean, and only time will tell if this barn of the recent past will become the model for barns of the future.

A photo and history of Robinson's Barn were provided by the Vermont Historic Preservation Commission in Montpelier. Although it is not especially old, being built in 1917, it has been nominated as a historic site.

This unusual ten-sided barn has a gable roof and is 1½ stories high. The ventilation cupola at the apex of the roof is also ten-sided, and twin silos with a gable-roofed addition complete the harmonious composition.

This beautifully kept barn is located in Strafford, Vermont.

This hexagonal barn, located in Newport, is architecturally unique in Maine, as far as can be determined. There are no other polygonal agricultural structures of any kind known to exist in the state, although some twenty octagonal houses have been identified so far.

What impulse led to the construction of this particular building is not known, nor has extensive research by the Maine Historic Preservation Commission revealed the name of the builder. It is presumed that the barn is contemporaneous with the ell to which it is attached and can be reasonably dated as *circa* 1850.

Aside from the door change to adapt it for use as a garage, the barn appears to have undergone no major alterations. Attached to the northwest side is a one-story frame ell with clapboard siding. This leads to the house, a frame Cape-style dwelling of the Greek Revival era.

This attractive old barn is gradually falling into complete disrepair. It is located in Orange, Vermont, and was built in 1906. Each wall is thirty feet high. The barn backs up against a bank on the eastern side, to which it is connected by a ramp, allowing vehicles to proceed directly into the hay-storage area. It has three floors and is topped with an octagonal cupola that matches the shape of the barn.

The Great Shaker Stone Round Barn in Hancock, Massachusetts, has stone walls three feet thick and is ninety feet in diameter. The original version of this barn was built in 1826, designed to take the place of several other community barns that had outlived their usefulness.

When that barn was put into use, it was learned that the internal circular ramp interfered with the main purpose of the barn, which was dairying; therefore, after the first barn burned in 1864, it was immediately rebuilt with the modifications necessary to improve its utility. The second version still stands today, a monument to the great stonemasons who built it. It is open to visitors, and it is really a sight to see both inside and out.

The influence of the Shakers on barn building in New England was great indeed; many a master barn builder traveled to see the Shaker barns and take notes on the innovations introduced in them.

Phil Ziegler 83

Restoration

Much has been said thus far that would indicate that many builders just pirate a few parts of an old barn and leave the rest to fall into a state of total ruin — an unfortunate practice. It is one thing to find a barn that is already on its way to oblivion and use what can be salvaged in a constructive way, quite another to tear out part of a perfectly preserved barn, such as the Val You Barn in Jericho, Vermont, and leave the rest weakened, vulnerable to further devastation and, finally, total collapse.

However, there is another side to the coin: the story of dedicated people who are saving these old structures, restoring them to their original greatness and beauty from foundation to rooftop. Some old barns are converted into unusual homes of beauty and charm. Others are reconstructed, either on their original sites or at new locations, where they are again as serviceable as they were when first built. Still others may become a part of a museum or educational complex, such as Wolf Trap in Vienna, Virginia.

Many of the drawings in this chapter are the result of the cooperation of Richard Babcock, staunch barn preservationist. His search for examples of the root barn — a barn found in its original setting that has not been improved or remodeled — is constant. The value of a root barn is based on two factors: First, it demonstrates the craftsmanship that went into the construction of the barn; second is the historical context, for these old barns were built and used during pre-Revolutionary times and the upheavals caused by the Civil War and Industrial Revolution, and thus provide priceless insights into the cultural development and growth of American agriculture.

In the South almost all of these root barns are gone forever, casualties of the ''scorched earth'' battles of the Civil War. It is in the North that such barns can be found, still standing on their original foundations.

This story involves two barns that were attached during restoration. The small barn in the foreground is thought to have been a blacksmith shop with its origins in the early 1700s. This is attested to by the presence of rough-hewn gunstock post beams, pole rafters with cross-pin fasteners, and the rigid center beam.

In 1808 this barn was moved to Shrewsbury, Massachusetts, between Worcester and Boston, and became a part of Harrington's Tavern and Horse Barn. Animals were rested in the barn while the travelers found refreshment in the parlor of the tavern.

In 1860 the Harringtons turned to dairy farming and erected the larger barn in the background, connecting it to the smaller barn. Recently, both barns were moved by Richard Babcock to their present location in Framingham, where they are an important part of the Macomber Farm and Education Center. The two barns now stand completely restored, reminders of the past in the bright and modern atmosphere of the Macomber Center.

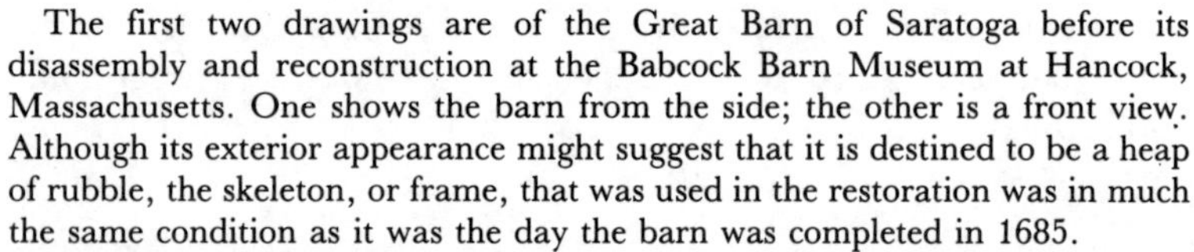

The first two drawings are of the Great Barn of Saratoga before its disassembly and reconstruction at the Babcock Barn Museum at Hancock, Massachusetts. One shows the barn from the side; the other is a front view. Although its exterior appearance might suggest that it is destined to be a heap of rubble, the skeleton, or frame, that was used in the restoration was in much the same condition as it was the day the barn was completed in 1685.

Jan Janse Bleeker, a Dutch immigrant who later became the seventh mayor of Albany, New York, built this barn to last. It was located within earshot of the Battle of Saratoga, and General Burgoyne and his troops sheltered themselves here. While many Indian attacks destroyed settlements throughout the area, this barn stood up to it all.

The Great Barn is unique in many ways. At fifty by fifty feet square, and thirty-five feet high at the peak, it was very large for the area. The roof has a pitch of ten over twelve, which means that for every foot of run there is a ten-inch rise in pitch. The main anchor beam was one foot wide, twenty-four inches deep, and over thirty feet long.

The wooden door hinges found in these old Dutch barns usually deteriorated quite rapidly, but those found in this barn were still in excellent condition. Because of the rarity of the find, one is shown in detail.

The last illustration shows the barn after it was completely reconstructed at the museum.

Phil
Ziegler
83

Phil
Ziegler
83

The Cider Barn was just exactly that, a barn housing an apple grinder and cider press. Built *circa* 1820 by Hiram Jones at Stillman's Village in Petersburg, New York, it is now a pile of rubble from which only a minimal amount of salvage could be used in the restoration of other barns. When located, it was discovered that about all that was holding up the barn were the apple grinder and the press. These were completely salvaged and moved to the Babcock Museum, where pure cider was made last year on the equipment, now on exhibit in the restored Great Saratoga Barn. The side beams and crossbeam of the press were so heavy that a small crane was needed to put the reassembled press in place in its new home.

The grinder, completely made of wood, is constructed so that either a horse or a couple of men can push the long handle in a circle, feeding the apples in the hopper at the front into the grinder. The mash comes out at the front into a container, which is then taken over to the press. A form placed on the press completely surrounds the runoff channels carved into the base plate. After the form is ready, a layer of clean straw is placed in the form, covering the bottom. Then a heavy layer of apple mash is put on the straw and, after moving the frame upward, the process is repeated until a mass about four feet high is formed. Then a heavy set of cross-hatched boards is placed on top of the mash pile and the screws are turned. The cider runs out the channel at the front into a container and is stored in kegs until used. Every once in a while the operator might put some of the mash aside and allow it to ferment, producing hard cider.

The construction of the Music Complex at Wolf Trap is an example of historic preservation and restoration of the highest order. Full credit for its conception must be given to Catherine Filene Shouse, whose dedication and love of antiquities made the whole project possible. Richard Babcock was the master builder on this project and one more skilled in the reconstruction of barns could not have been found. Only the tools available to the original settlers were used at Wolf Trap; even the clothing worn by the workers was in the style of Colonial times.

Barns of certain specifications were needed for this major project: most important, one with a threshing-floor area large enough to seat an audience of three hundred people. A German barn of sufficient dimensions was found along the flat-lands of the Schoharie River near Blenheim, New York. Although it had outlived its usefulness as a working barn, it was perfect for the concert-hall section of the new building. All of the beams in this German-style barn were hand hewn, and the braces and boards show evidence that they were produced by an up-and-down saw. It was built in the early 1800s and is forty feet wide and fifty feet long, with side walls of twenty feet. The top of the gable roof measures thirty feet from the ground.

Originally, the barn was the property of Harold and Gertrude Kniskern, who were the first settlers to make a living off the land in this area. Both were buried near the site where the barn originally stood.

The second illustration in this series of four is a Scottish barn built about 1791. It is a three-bay, medium-sized barn measuring thirty by forty feet. This first-generation barn was built much like those of the builders' ancestors in Scotland, but with an important difference: Because the trees in New England were so much bigger than those found in Scotland, the beams in this barn were much larger. This barn originally stood beside Route 22 in Jackson, New York.

Eventually, the Scottish-style barn was moved to a hillside site and, taking advantage of this location, the owner added a lower level to accommodate livestock while the main floor was used for hay storage. A lean-to was eventually attached to the back, and later a slate roof was added. The timbers were all hand hewn and the side girts and knee braces were of up-and-down sawn oak.

In this barn an ingenious method of securing the siding boards was developed because of the scarcity of nails at the time. The plates and the sills were grooved in such a way that the boards could be slipped right into the grooves and were thus anchored for good. When the slate roof was added, extra supports had to be installed to support the additional weight. The sills and the plates, all of one piece, ran the entire length of the barn, and the builder probably used the bull wheel and gin pole on this job. The third drawing shows the smaller barn after its restoration.

When these two old barns were reconstructed into one large building and the finish work done, the music building at Wolf Trap was complete, with its modern exterior in harmonious counterpoint to the faithfully restored interior. The final drawing shows Wolf Trap as it stands today, a monument to those whose ingenuity, skill, and determination made the building of the old original barns possible. The barn at the far left is the German barn and on the far right, the Scottish barn.

At the Sleepy Hollow restoration site known as Phillipsburg Manor in Tarrytown, New York, a Dutch barn had been restored as part of the complex. It was completely destroyed by fire in August 1981, but Sleepy Hollow officials decided to replace the barn if one could be found. Again, Richard Babcock was consulted, and he soon located a replacement that met all specifications.

This Dutch barn, built between 1720 and 1750, was originally on the manor of Steven Van Renssalaer near Guildersland, New York. It had been leased to David Ogsbury and his wife Elizabeth, who soon purchased the property outright. However, unlike many farmers who had purchased such land from Renssalaer, Ogsbury kept some of the slaves and even admitted to having sired children with a slave named Gin.

Phil
Ziegler
83

John Smedley purchased the land upon which the Moon Barn was eventually built in 1770. Although he cleared the land for farming, thus increasing its resale value, when he sold it in 1825 it was still devoid of buildings. About this time the Erie Canal was opened, and many farmers began to migrate to the more fertile land of the Midwest from the rocky soil of New England.

The farm was purchased by Alfred Moon in 1880. This barn had been built in the 1850s and was still standing on its original site until 1975.

During the 1976 bicentennial celebration, the barn was moved to Williamstown, Massachusetts, across from the Rosenburg Center at Hopkins Forest, where it stands today, restored as the result of intense community cooperation. (Richard Babcock has written a beautiful pamphlet describing this project from beginning to end.) The barn now serves as the headquarters for the Hopkins Forest Farm Museum, which is dedicated to the agricultural heritage of the Berkshires. The following two illustrations show the barn as it stood in its root state and after its restoration.

Conclusion

Try taking a side trip sometime up one of those narrow, seemingly deserted dirt roads that we all pass every day in our haste to go nowhere. You might just be on the verge of taking a trip into the past and to an unforgettable discovery.

It is my most sincere hope that this book will provide the impetus for you, dear reader, to turn off that boob tube one of these fine days and go look for an old barn — and don't forget to take the kids along. Soon such opportunities may be gone for good.

Writing and illustrating this book has been a great adventure. Many times my memory wandered back to the years when as a young boy I was "turned out to pasture" by my folks each summer on my grandparents' Ohio farm. Although it was just a small homestead, it had chickens, livestock, a big chestnut tree in the yard, and, of course, a barn. Even then, I was in love with barns.

I can still smell the sweet-scented new hay in the loft, where I used to lie watching the tiny motes of dust that floated in the shaft of sunlight piercing a small hole in the roof. In my mind's eye, the chickens are still there, pecking at some delicacy as they stalk and scratch in the yard below. In memory, the swallows still flit around the rafters and hand-hewn beams. I can still feel the texture of warm, weather-beaten boards, whose beauty, like that of some people, is only enhanced by age.

Is it not our *duty* to help preserve these fine, historic buildings whose integrity of form and function result in such understated beauty?

Bibliography

The American Agriculturist, 23, New York: Orange Judd Co., 1874.

Arthur, Eric, and Dudley Whitney. *Barns.* Ontario: M.F. Fehelen Arts Co., Ltd., 1972.

Beard, Frank, and Bette A. Smith. *Maine's Historic Places.* Camden, Me.: Down East, 1982.

Barn Plans and Outbuildings. Orange Judd Co., 1897.

Congdon, H.W. *Old Vermont Houses.* New York: Alfred A. Knopf, 1946.

Dickerman, Charles W. *The Farmer's Book.* Philadelphia: Ziegler, McCurdy, and Co., 1869.

Dodds, Eugene. *The Round Stone Barn.* Shaker Community, Inc., 1968.

Foley, Mary Mix. "The American Barn." *Architectural Forum* (August 1951).

Garvan, Anthony. *Architectural and Town Planning in Colonial Connecticut.* New Haven: Yale University Press, 1961.

Glassic, Henry. *Pattern in the Material Folk Culture of the Eastern United States.* Philadelphia: University of Pennsylvania Press, 1969.

Hedrick, Ulysses Prentice. *A History of Agriculture in the State of New York.* New York: Hill and Wang, 1966.

Hill, Ralph N., Murray Hoyt, and Walter R. Hard, Jr. *Vermont, A Special World, Sixth Edition.* Boston: Houghton Mifflin Co., 1979.

Kimball, Sidney Fisk. *Domestic Agriculture in the American Colonies and the Early Republic.* New York: Scibners, 1973.

Klamkin, Charles. *Barns.* New York: Hawthorne Books, Inc., 1973.

Konrad, Victor A. "Against the Tide: French Canadian Barn Building Traditions in the St. John Valley of Maine." *The American Review of Canadian Studies,* 12, no. 2, 1982.

State O' Maine Facts, Fifteenth Edition. Camden, Me.: Down East, 1982.

Tunis, Edwin. *Colonial Craftsmen.* New York: Thomas Y. Crowell Co., 1965.